MUSHROOM CULTIVATION TECHNOLOGY

MUSHROOM CULTIVATION TECHNOLOGY

ROBIN GOGOI
YELLA RATHAIAH
TASVINA RAHMAN BORAH

SCIENTIFIC PUBLISHERS

Published by:
Scientific Publishers
5A, New Pali Road, P.O. Box 91
Jodhpur 342 001, India
E-mail: info@scientificpub.com
Website: www.scientificpub.com

Print : 2020

ISBN: 978-81-72334-18-5
eISBN: 978-93-87893-01-6

Printed in India

PREFACE

Mushroom is the God given natural health food for mankind. Many people are now eager to know about the skills for growing mushrooms on a small scale. Anyone, who wants to start cultivation of mushroom, will search for a simple and practical oriented book on mushroom. This book deals mainly with the practical aspects of mushroom production. All the chapters in this book are based on personal experience of the authors during the last ten years.

Spawn, i.e. the mushroom seed is the key input for mushroom cultivation. Therefore, the spawn production technology is described in a simple language. Further, cultivation of four types of mushrooms is dealt with. Our focus is on how to grow these mushrooms under the prevailing climatic conditions, i.e. seasonal cultivation.

Now-a-days, many people are interested in growing milky mushroom, because it is relatively new. Its method of cultivation is also described. A chapter on how to recognize common wild edible mushrooms found in Assam is also included. This book will be of use for Agricultural students who have taken course on mushroom cultivation. Besides, this will also serve as a guidebook for entrepreneurship development on mushroom.

We are highly indebted to Dr. P.C. Deka, Former Dean, Faculty of Agriculture and present Vice Chancellor of Tezpur University for his constant encouragement to write this book. We are grateful also to Dr. A.K. Phookan, Professor and Head, Department of Plant Pathology for providing us facilities to conduct works on mushroom research.

Robin Gogoi
Senior Scientist, AICRP on Potato,
Department of Plant Pathology of
College of Agriculture,
Assam Agricultural University, Jorhat

Yella Rathaiah
Professor, Department of Plant Pathology,
College of Agriculture,
Assam agricultural University, Jorhat

Tasvina Rahman Borah
Research Associate,
Department of Plant Pathology,
Assam Agricultural University, Jorhat

CONTENTS

CHAPTER I

Introduction

1.1. WHAT IS MUSHROOM?

Mushrooms are the eucaryotic, spore bearing organisms under macro-fungi, lacking chlorophyll and grow on dead decomposed matter as saprophytes. They derive nutrients through their mycelia.

The term 'Mushroom' is derived from many languages – (i) Latin 'fungo' means to flourish, i.e. growing out from the ground or from trees. (ii) Greek 'sphonggos' or 'sphoggos' means 'sponge', i.e. species having sponge like structure and (iii) French 'mousseron' (muceron), 'mousse' or 'moss'. Other variations are 'muscheron' and 'mouscheron'.

Mushroom has got different names in different languages, viz. '*ksumpa*' in Sanskrit, '*khumbi* or *khumb*' for button mushroom in Hindi, '*dhingri*' for oyster mushroom in Hindi, '*kathphula*' in Assamese. Other words in Sanskrit are *chatra* for the fleshy-capped fungi, *bhoomi kavak*, *kavaka*, *bhustrna* and *kukurmutta*.

1.2. TOADSTOOLS

Poisonous mushrooms are called as the toadstools. Some of the important examples are *Amanita phalloides* (The death cap), *A. muscaria* (Fly agaric) and *A verna.* (Destroying angel).

1.3. SPECIES OF EDIBLE MUSHROOM

Four extensively grown edible mushroom are – (i) Button mushroom (*Agaricus* spp.), also called as White Button, European button, Temperate mushroom, (ii) Paddy straw mushroom (*Volvariella* spp.), also as Chinese mushroom or Tropical mushroom, (iii) Oyster Mushroom (*Pleurotus* spp.), (iv) Milky mushroom (*Calocybe indica*) and (v) Shiitake (i.e. Oak tree in Japanese) mushroom (*Lentinula edodes*). The species grown to some extent are-Black ear (*Auricularia auricula*), Giant fungus (*Stropharia rugosa-annulata*), Enokitake (*Flammulina velutipes*), maitake (*Grifola frondosa*), Jelly fungus (*Tremella fuciformis*), Nameko (*Pholiota nameko*) and Truffle (*Tuber melonosporum*).

- Species of Button Mushroom: *A. bisporus* and *A. bitorquis*.
- Species of Paddy straw mushroom: *V. volvacea*, *V. diplasia* and *V. esculenta*.
- Species of Oyster mushroom: *P. citrinopileatus*, *P. cornucopie*, *P. flabellatus*, *P. florida*, *P. eous* (pink mushroom), *P. ostreatus* (Hirakitake), *P. sapidus* and *P. sajor-caju* (grey oyster).

1.4. POISONOUS vs. EDIBLE MUSHROOMS

Some of the morphological features should be considered to distinguish these two groups of mushrooms (Fig. 1).

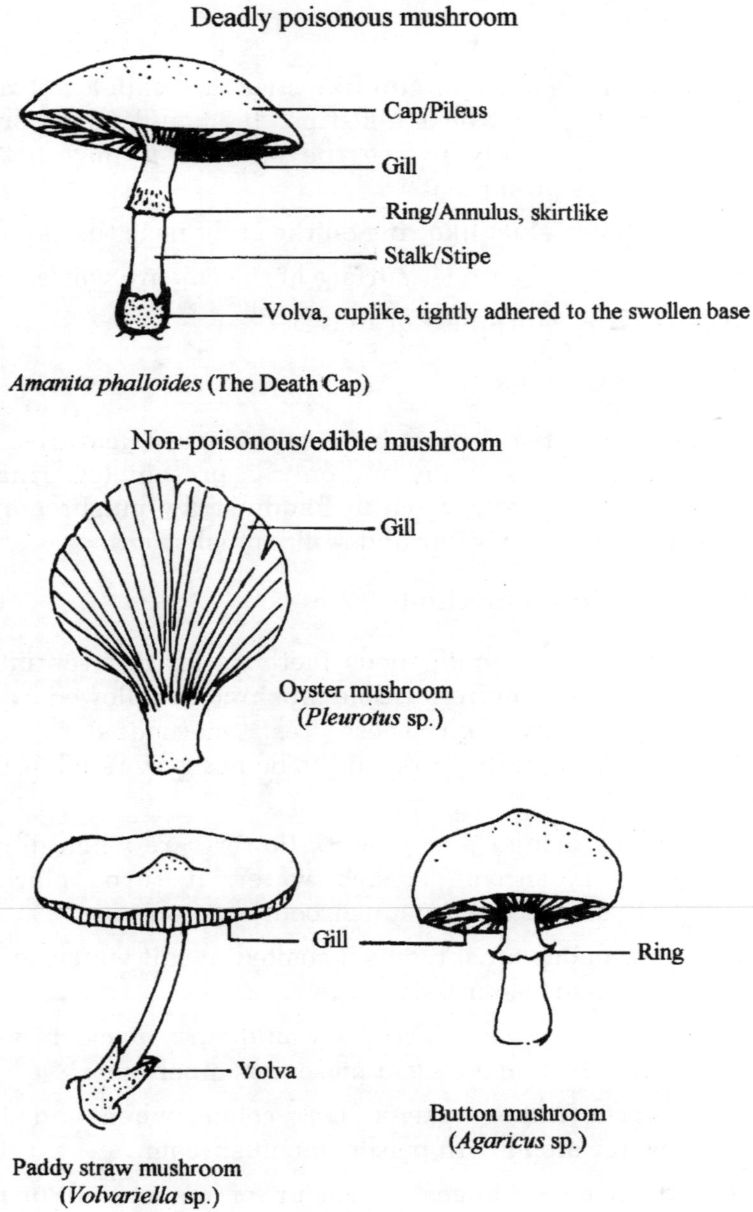

Fig. 1. Poisonous and Edible Mushroom

Death cap

i. There is a sac or cup like structure called 'volva' at the base of the stalk/stipe. It should be observed very carefully to see the skin as it may not be always prominent.

ii. It has a skirt like ring on the stalk near the cap.

iii. Gills on the under surface of the cap are white.

iv. It grows up singly or in twos.

Edible mushrooms

Out of the two important morphological features viz. volva and ring, either any one only is present (eg. ring in Button mushroom and volva in Paddy straw mushroom) or none is present (eg. Oyster and Milky mushroom).

1.4.1. Traditional methods

There are some indigenous methods for differentiating poisonous mushroom from edible mushrooms followed by the local people. Although these test methods are very commonly followed, they are said to be not very reliable. The methods are:

i. Some brinjal pieces or its flowers are when dipped into mushroom soaked water, it turns blackish which indicate the mushroom is poisonous.

ii. Poisonous mushrooms if soaked in salt water, give a blackish colour to the water.

iii. Silver or copper articles would turn blackish when immersed in a dish of poisonous mushroom.

iv. Garlic flakes take a dark colour when added to water along with poisonous mushroom.

v. Unusual changes in colour of onion and brinjal occur while cooking along with poisonous mushroom.

vi. Milk curdles immediately in contact with poisonous mushroom.

1.4.2. Other features of poisonous mushroom

i. The poisonous mushrooms are comparatively soft and the skin of the pileus cannot be easily peeled off.

ii. Poisonous mushrooms are generally colourful and quite attractive.

iii. Exudation of milky substances is seen from damaged fruit bodies of some poisonous mushroom species.

iv. Poisonous mushrooms are usually bitter or sour in taste and bear unpleasant smell.

v. Generally ants, insects and other animals do not feed on poisonous mushrooms.

1.5. NUTRITIVE AND MEDICINAL VALUES OF MUSHROOMS

1.5.1. Nutritive values

i. Protein: The protein content of mushroom is 2.5-3% on fresh weight basis and 20-35% on dry weight basis. The mushroom protein is of good quality and its digestibility is high.

   Cereals are deficient in two essential amino acids, *viz.* lysine and tryptophan. But mushrooms are rich in these two amino acids apart from most of the essential amino acids *viz.* leucine, isoleucine, valine, threonine, tyrosine and phenylalanine which human body can not synthesize. However, two other amino acids, viz. methionine and cysteine are less in mushrooms but they are present in cereals. Therefore, mushrooms and cereal diets complement each other.

ii. Vitamins: Mushrooms are rich in vitamin B complex and vitamin C. Under B complex group, thiamine, riboflavin and niacin are present in good amount. Another plus point is that vitamins in mushrooms are not lost during cooking. Mushrooms also contain folic acid and vitamin B12 which are absent in green vegetables.

iii. Minerals: Mushroom possess a high amount of potassium. Besides, the ratio of potassium:sodium (K:Na) is very high in mushroom. This is an advantage for persons having high blood pressure or heart problem.

iv. Carbohydrate: The total carbohydrate content in mushroom is 4-5%. It comprises of chitin, hemicallulose and glycogen. Mushrooms have very little quantity of sugar (0.5%).

v. Starch is absent.

1.5.2. Advantages of mushroom as diet

i. Low in calories, i.e. 32 Kcal/100g fresh mushroom and low in fat (maximum 0.3%). Therefore, mushrooms are suitable for obese persons.

ii. Cholesterol is absent, but ergosterol is present which is converted into vitamin D in human body. Therefore, mushrooms are good for heart disease.

iii. High fibre content, i.e. above 1%. Therefore, it is suitable for those having constipation.

iv. Mushrooms are alkaline due to alkaline ash (more than 1%). Therefore, it is an advantage for those having acidity/gastritis problem.

1.5.3. Therapeutic uses

i. Ability to reduce blood cholesterol: Mushrooms possess specific blood cholesterol reducing substances. For example, *Agaricus bisporus* and *Lentinula edodes* can bring 34 and 35% decrease in blood cholesterol, respectively.

ii. Hypoglycemic effect: *Coprinus comatus* has antidiabetic effect. Oyster mushroom is an ideal diet for diabetic and obese persons since carbohydrate and fat are extremely low.

iii. Antitumour effect: Some mushrooms have anticancerous effect. *L. edodes*, *P. ostreatus* and *A. bisporus* can suppress tumour growth up to 40%.

iv. Antiviral effect: A crude extract from shiitake mushroom can regress the influenza and polio viruses. Some species are known to be effective against HIV of AIDS also.

1.6. IMPORTANT POINTS

i. There is no chance of poisoning if the cultivated species like oyster mushroom, paddy straw mushroom, milky mushroom and button mushroom are used as food item.

ii. Wild mushrooms should not be consumed without proper identification or knowledge.

iii. Liquor should be avoided along with the mushroom dish.

iv. Pumpkin or broad beans, when consumed along with mushrooms can cause food poisoning.

CHAPTER II

Chronological Events on Edible Fungi

2.1. EVENTS OF PREHISTORIC ERA

Aryans role: Around 1500 B.C., the Aryans migrated into the Indian subcontinent who used an intoxicating drink *soma* in their religious rites.

Rig Veda : There are many songs on such *soma* in the *Rig Veda* that refers to *Amanita muscaria* (Wasson, 1969).

Stone carving : O'Heer and his associates (1886) found traces of puffballs and many other fungi carved on some stones.

Mayan period : Mushroom shaped stone carvings were found in Central America and some highlands of Guatemala. These were assumed to be used in their many ceremonies. Mushroom shaped objects resembling to *Amanita muscaria* were figured in the astrological paintings of Mayancodices. During this period, various names of mushroom fungi were given e.g. *Agarikon, Amanita, Bolites* etc.

Roman period : The Romans were very selective, elite and artistic in their food habits. They felt a feast is complete and luxurious on serving different mushroom preparations. Poet Horace designated the best to 'Pratanses', a field mushroom, among all fungi. Pliny, the Roman naturalist, illustrated two most delicious mushroom namely *Amanita caesaria* and *Boletus edulis* and noted about their recipes which were served in the special occasions to the masters and princes only in attractive and costly utensils.

2.2. CULTIVATION OF MUSHROOM

A. Button Mushroom

1638-1715 This is the period of Louis XIV where cultivation of mushrooms was recorded for the first time.

1700 As recorded by Treschow, before this period, there was a fashion of cultivating melon and pineapple in hot beds of compost under heat where *A. bisporus* appeared spontaneously. This was observed for several times that gave the idea of growing button mushroom.

1707 At Royal Academy of Science, France, Tournefort, mentioned the earliest account of compost preparation and mushroom cultivation technology and it was published in Paris.

1731 Miller introduced French method of cultivation of button mushroom into England.

1754 Lundberg of Sweden described about growing mushroom in greenhouses. This is the origin of growing mushroom above ground.

1779 Abercrombie described a method of composting stable manure in stacks.

1800-1810 A French gardener Chambry started growing of button mushrooms underground in the quarries in Paris. He kept horse manure in a big heap and allowed to heat up naturally that finally converted into compost. The soil of the horse-grazing field where edible mushrooms were naturally growing was used as spawn to inoculate the prepared compost. Production of mushroom by this method was very sporadic and just a matter of chance.

1831 Callow reported the mushroom house introduced by Oldacre (a gardener to the late Sir Joseph Banks) that could be utilized for growing mushroom throughout the year. Moreover, in England, Callow illustrated for the first time about the mushroom house designed with many shelves and flat beds. He grew mushroom in cropping house warmed by fire heat by which he could get fairly good yield.

1893 Costantin pointed out that the incidence of diseases made continuous changing of growing area.

1894 Matrochot and Costantin among the French mycologists could overcome the problem of spawn preparation artificially. They could germinate spores of mushroom fungus, prepared it pure culture and used successfully to prepare spawn. Although this method was patented but not commercialized. Further, both mycologists solved another problem of mushroom cultivation. They found out the cause of the *la mole* disease of mushroom as *Mycogone perniciosa* and also controlled by fumigating with sulphur.

1902 Miss Fergussen of Corne, America numerated the detail of the spore germinating method and growing of mycelia.

1905 In America, Dugger perfected the method of spawn preparation and produced pure culture spawn from tissue of mushroom caps. This technology helped in setting up of number of American spawn companies.

1915 A second stage of composting called sweating out was introduced by mushroom growers in USA. Composts were allowed to complete thermogenesis shelved compost beds within the growing house before spawning.

1918 Pure culture spawn introduced by Lambert was marketed in the United State. This spawn was prepared in the milk bottles containing sterilized horse manure compost and plugged with cotton wool. In this year, European countries also started spawn production commercially (Circa, 1918).

1929 Lambert discovered that productive spawn could also be prepared from single spore cultures that helped further improvement of strains.

1932 Sinden patented his grain spawn method.

1936 Pizer, a French at Wye College first discovered that the growth of spawn improved tremendously in the compost when ground gypsum (calcium sulphate) was used during composting. Gypsum prevented greasiness in the compost.

1937 Sinden found that about one third of monospore cultures of *A. bisporus* were incapable of producing fruit bodies.

1941 Lambert identified the main regions in the compost stack where a temperature of 50-60^{0}C with adequate aeration is maintained can yield good quality compost.

1950-1953 Sinden and Hauser introduced Short Method of composting.

1965 Block introduced container system of pasteurization in which first part of outdoor composting was done for 10 days followed by 8-10 days treatment in a container or bin.

1971 A hybridization method for *A. bisporus* was introduced by Robert Miller.

1972 Randle and Hayes composted in insulated metal drums supported on motorized rollers for mixing of contents.

1973 Somycel, a French Firm commercially introduced the first strain of *A. bitorquis* as strain No. 2017 and later by Le Lion.

1981 Eicker devised an insulated and ventilated bin for composting small amount of material.

1983 Smith reported the preparation reported short duration of mushroom compost without the use of steam.

1988 Leendertsee developed indoor composting systems to overcome some problems in outdoor composting.

1990 Gerrits and Griensven reported the new developments in indoor composting (tunnel process).

B. Oyster mushroom

1917 Falck first described successful cultivation of *Pleurotus ostreatus* on tree stumps and logs.

1951 Lohwag first grew *Pleurotus* on saw dust mixtures.

1959 Three scientists namely Block, Tsao and Hau grew large number of fruit bodies on sterile saw dust/oat meal mixtures.

1962 Bano and Srivastava reported mass production on straw-based substrates and their work gave the right way for large scale commercial exploitation. They obtained increased yield by using paddy straw.

C. Paddy straw mushroom

1934 In Burma, Baker gave detail account of the cultivation of *Volvariella*.

1939 The Agriculture Department, Chennai (then Madras) undertook cultivation of paddy straw mushroom (*Volvariella volvacea*) and also repeated in 1945. But the work was discontinued later.

1940 Su and Seth published the procedure of spawn production of paddy straw mushroom.

1943 In Madras, Thomas *et al.* gave the complete method of cultivation of paddy straw mushroom (*V. diplasia*).

1947 Asthana first used red gram dal powder to the beds of paddy straw mushroom by which he obtained more yield. He determined the best period of growing this mushroom as April to June in Central Provinces and also carried out the chemical analysis of this mushroom.

2.3. MAJOR ACHIEVEMENTS IN INDIA

1886 In India, N. W. Newton first grew some specimens of edible mushroom and exhibited at

	the Annual Show of Agriculture, Horticultural Society of India.
1996-1997	Dr. B.C. Roy of the Kolkata (then Calcutta) Medical College conducted some experiments on chemical analysis of the local mushrooms collected from caves and mines.
1908	Sir David Prain was much interested to the mushrooms used by the poor as food in famine stricken areas. Hence he instituted a thorough survey on the edible mushrooms. Its detail was published in the Indian Science Congress held at Nagpur in 1926.
1921	S. R. Bose successfully cultured two agarics on a sterilized dung medium.
1940	S.R. Bose and A.B. Bose enumerated the method of growing mushrooms on horse manure.
1941	Padwick analyzed and pointed out the possibilities and obstacles in cultivation of *A. bisporus* as its cultivation in most of the time remained unsuccessful in the country during which it was grown very successfully throughout the world.
1961	A scheme on 'Development of Mushroom Cultivation in Himachal Pardesh' was started by the Department of Agricutlture, Himachal Pradesh in collaboration with the Indian Council of Agricultural Research (ICAR). This was the first attempt on cultivation of *A. bisporus* in the country. Cowdung was used for cultivation of button mushroom that did not give good yield.
1964	Horse manure compost was used by the same department where only button mushroom could be cultivated successfully. On experimental

basis, the Council of Scientific and Industrial Research (CSIR) and State Govt. at Srinagar in J & K started cultivation of *A. bisporus*.

1965 Dr. E.F.K. Mantel, F.A.O., Mushroom Expert, guided and assisted Deptt. of Agriculture for construction of a modern spawn laboratory and a completely air-conditioned mushroom house. Other aspects of research were evaluation of different strains and use of various agricultural wastes and organic manures and fertilizers for preparing synthetic compost. Dr. Mantel's consultancy was over after 7 years.

1974 Dr. W.A. Hayes, F.A.O., Mushroom Expert, guided to further improvement of the compost preparation method, pasteurization and management of important parameters in the mushroom house. He standardized new compost formulations, casing materials and important parameters like nitrogen content in the compost, moisture in the casing soil, air movement and maintenance of proper environmental factors that increased mushroom yield tremendously.

1977 The Department of Horticulture (H.P) launched Mushroom Development Project under the UNDP. Mr. James Tunney rendered his expertise to this project. He got bulk pasteurization chamber constructed and made readymade compost and casing soil available to the growers of Himachal Pradesh. The project was concluded in 1982 and now it has been running by the Department of Horticulture (H.P.). Dr. R.L. Munjal in Dr. Y.S. Parmar University of Horticulture and Forestry, Solan made significant contributions viz. compost formulations, use of spent compost as casing material and management of important pests and diseases.

1982 The ICAR sanctioned the creation of National Centre for Mushroom Research and Training (NCMRT) at Solan during VI Plan on October 23, 1982 with the objective of conducting research on problems of mushroom production, preservation and utilization and to impart training to scientists, teachers, extension workers and interested mushroom growers.

1983 NCMRT (presently National Research Centre for Mushroom, NRCM from 1999) started functioning from June 8, 1983. The All India Coordinated Mushroom Improvement Project (AICMIP) was sanctioned by the ICAR during VI Plan w.e.f. April 1, 1983 for two years with head quarters at NCMRT. Initially six centers were sanctioned under this project. Those are located at GBPU of Agriculture and Technology, Pantnagar (Uttaranchal), Punjab Agricultural University, Ludhiana, Tamil Nadu Agricultural University, Coimbatore, Bidhan Chandra Krishi Viswa Vidyalaya, Kalyani, (WB), Mahatma Phule Agricultural University, College of Agriculture, Pune (MS) and C.S. Azad University of Agriculture and Technology, Kanpur (UP). Later, Kalyani and Kanpur Centers were withdrawn and Indira Gandhi Krishi Vishwa Vidyalaya, Raipur (MP) centre was added.

CHAPTER III

Production of Spawn

Spawn, i.e. seed of mushroom consists of mycelium of the fungus grown on suitable substrates like cereal grains. It is the key input of mushroom cultivation like any other crop cultivation using their seeds. As the mycelia of mushroom fungus as such is not possible to propagate or mix into the main substrates of cultivation like straw, hence the mycelia are carried or multiplied on a carrier like cereal grains. Non-availability of spawn is the major constraint in mushroom production. Spawn should have the following standards:

i. a high yield potential

ii. cost should be economical

iii. substrate for spawn should be easily available.

The following steps followed to prepare spawn that are mentioned below.

3.1. PURE CULTURE FOR SPAWN PREPARATION

Pure culture of edible fungi is made by tissue culture technique on a culture medium like potato dextrose agar

(PDA) or oat meal agar (OMA) so that it can be used for preparation of spawn. Malt extract medium (25g malt extract + 20g agar agar are dissolved in 1 l of water) is the best medium for culturing button mushroom.

3.2. PREPARATION OF POTATO DEXTROSE AGAR (PDA) MEDIUM

PDA is the best medium for growing the culture of oyster and Paddy straw mushrooms.

Formula:	Potato	250 g
	Dextrose or glucose	20 g
	Agar-agar	20 g
	Distilled water	1 litre

Method

a. Wash 250 g of potatoes, peel off the skin with a sharp knife. Cut potatoes into pieces.

b. Put the potato pieces into 500 ml of water in a saucepan with lid and boil for 20 minutes or until potatoes become soft. Alternatively, cook the potatoes by steaming for 15 minutes in a pressure cooker or an autoclave.

c. Melt the agar-agar in another 500 ml of water, in one litre flask. This should be melted in a water bath or in a pressure cooker. The agar may spill out when the intensity of heat is high. Therefore, close the mouth of the flask with a cotton plug. For saving time and electricity, both potatoes and agar-agar can be heated at a time in a pressure cooker or an autoclave by steaming for 15 minutes.

d. Strain and squeeze the juice from the cooked potatoes through a muslin cloth.

e. Add 20 g of dextrose or glucose to the potato extract.

f. Mix the molten agar and potato extract. Make up the volume to 1 litre by adding warm water.

g. Check the pH with the help of pH strips.

h. Adjust the pH at 6 by adding drops of 0.1 N HCL or 0.1 N NaOH. If the pH is 4 or below, the agar does not solidify after autoclaving.

i. Pour the medium into culture tubes or 100 ml conical flasks. Wipe off the media sticking to the mouth of the tubes and flasks.

j. Put the cotton plugs; cover them with pieces of butter paper.

k. Allow the steam to go out for 15 min, then close the steam outlet of the autoclave. This step is a must.

l. Sterilize the medium in an autoclave or pressure cooker at 15 lbs pressure/sq. inch for 30 minutes. Do not allow the pressure go beyond 15 lbs i.e 121°C. At higher temperature dextrose may change its property. The carbohydrates are partially hydrolyzed at high temperature. Further, excessive autoclaving partially hydrolyses agar which can inhibit growth of fungi.

m. It is important to add antibiotic to PDA in order to check growth of bacterial contaminants coming from mushroom tissues. Antibiotic is destroyed by heating, Therefore it is added after autoclaving. Take out the PDA from the autoclave after sterilization and keep it in the laminar flow room. Add a small quantity (2 mg) with the help of inoculation needle to PDA just prior to its solidification. Stir the medium and make the slants. The PDA slants should be wrapped in a polypropylene bag and stored in a refrigerator.

How to prepare 0.1 N solution of hydrochloric acid?

Take 8.4 ml of concentrated hydrochloric acid and add to 992 ml distilled water. Store the solution in a bottle with label.

How to prepare 0.1 N solution of sodium hydroxide?

Dissolve 4.2 g of sodium hydroxide in one liter of distilled water. Store the solution in a bottle with label.

3.3. PREPARATION OF MOTHER CULTURE

First generation (T_1) pure culture prepared from the tissue of mushroom fruit body (sporophore) is called mother culture. The strength of mushroom culture is maintained by preparing mother culture. The culture becomes weak when subcultured several times.

Method (Fig. 2)

a. Select a fresh, big size fruit body of the mushroom. It should not contain much of water.

b. Split the mushroom longitudinally into two equal halves in a laminar air flow cabinet.

c. With a new blade, cut a cube of tissue 3 mm x 6 mm from the inner side of the junction of cap and stalk.

d. Place the tissue on PDA slant or plate and incubate at 25°C for 10 days. The antibiotic Streptopenicillin or Streptomycin should earlier be added to PDA.

   After 48 hrs, the block of tissue turns into powdery white cottony growth indicating success of the culture.

e. Store the flasks containing pure culture of the mushroom in a refrigerator when the cottony mycelium covers the entire PDA surface.

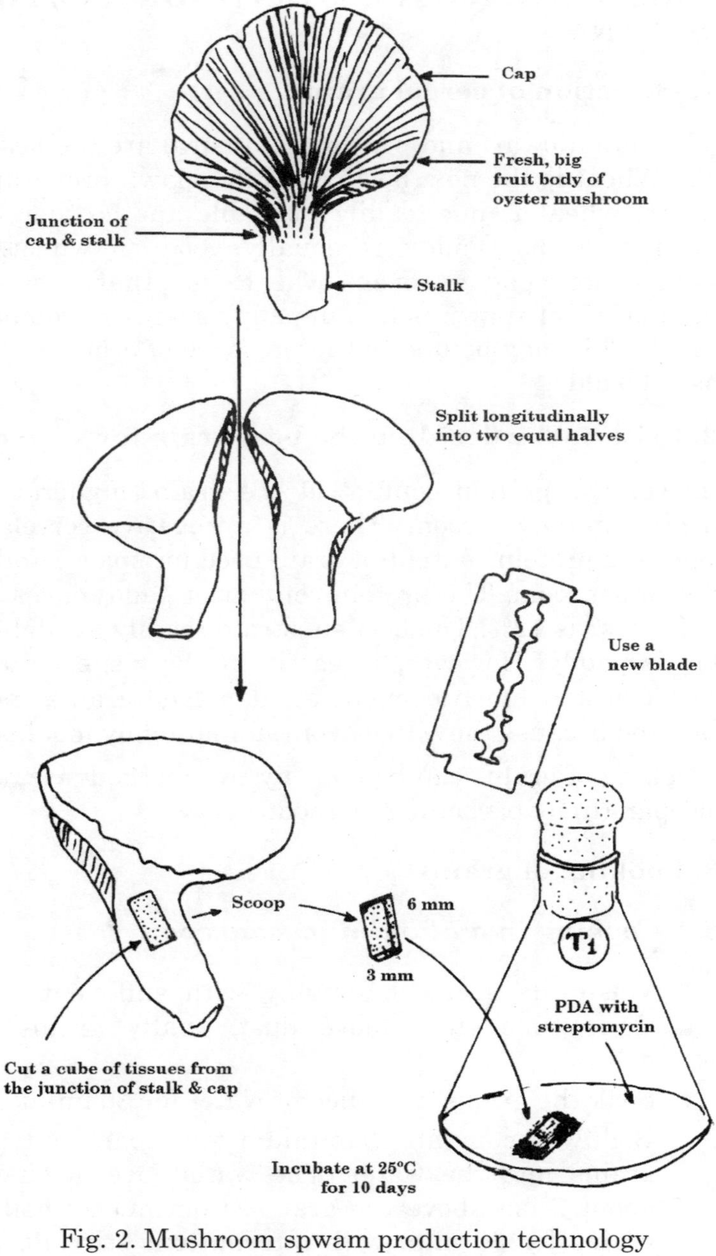

Fig. 2. Mushroom spwam production technology

3.4. PREPARATION OF SUBSTRATE (GRAINS) FOR SPAWN

3.4.1. Selection of cereal grains

Sorghum (jowar) and pearlmillet (bajra) are the best for spawn. Wheat is the next best cereal for spawn preparation. In Assam wheat is not readily available and costlier than the paddy grains. Paddy is equally good as wheat. An advantage of paddy over wheat is that bacterial contamination of spawn is less in paddy grains as compared to wheat. This may be due to the presence of a harder husk in case of paddy.

3.4.2. Criteria for deciding the best grain for spawn

Lower the protein content of the grain, higher is the yield of oyster mushroom. There is a negative correlation between the protein content of grain used for spawn and the yield of mushroom. The protein content of paddy is 7% and that of wheat is 12%. The best substrate readily available in Assam is paddy. However, wheat is the best substrate for spawn of button mushroom. Insect damaged grains should not be used because they invite contamination by moulds.

Cooking of grains can be done by two methods viz. (i) in a saucepan (ii) by parboiling method.

3.4.3. Cooking of grains

3.4.3.1. Cooking in a aluminium saucepan

a. Wash the grain thoroughly with sufficient water three times to remove dust, chaffy grains and debris.
b. Soak the grains in sufficient water for 30 minutes.
c. Boil water in an aluminium saucepan and pour grains into the water. The water level should be about 2 cm above the grains. Continue to boil the grains for 30 min. with intermittent stirring. Test

for softening of the cooked grains by pressing between fingers. Stop boiling when very few grains show splitting. Over boiling will result in complete splitting of grains and stickiness.

d. Drain off excess water by spreading the cooked grains on a sieve made of wire mesh.

e. The grains are spread thinly on floor for 1hr so that the water on the surface of grain is evaporated.

f. Mix thoroughly 20g of good quality calcium carbonate ($CaCO_3$) with one kg of cooked grain. The purpose of coating the grains with $CaCO_3$ is to raise the pH. It should be 8.5 or above, before autoclaving. The pharmaceutical grade $CaCO_3$ is the best as well as economical. The growth of the spawn is fast when the pH of the grain is optimum, i.e. above 7.5. Growth of spawn is slow, when the pH is 6. The cheaper grade of $CaCO_3$ viz. French chalk does not raise the pH to desired level. In consequence, the competitor moulds viz. *Trichoderma* and *Penicillium* will dominate. A finer powder of $CaCO_3$ gives the best coating on the grains when compared to a coarse powder.

3.4.3.2. Cooking by parboiling method

Parboiling of paddy is a common practice in Assam. A mini parboiling drum (Fig. 3 and Fig. 4) is used for spawn preparation. With this equipment, both soaking and cooking of paddy is completed in the same container in 5½ hrs. The capacity of the drum is 11 kg. paddy.

Method

a. Fill the parboiler with 25 l of water.

b. Heat it over fire until the temperature of water reaches 85°C. Pour 10 kg of washed and cleaned paddy grains into the hot water.

Fig. 3. Mini parboiling drum

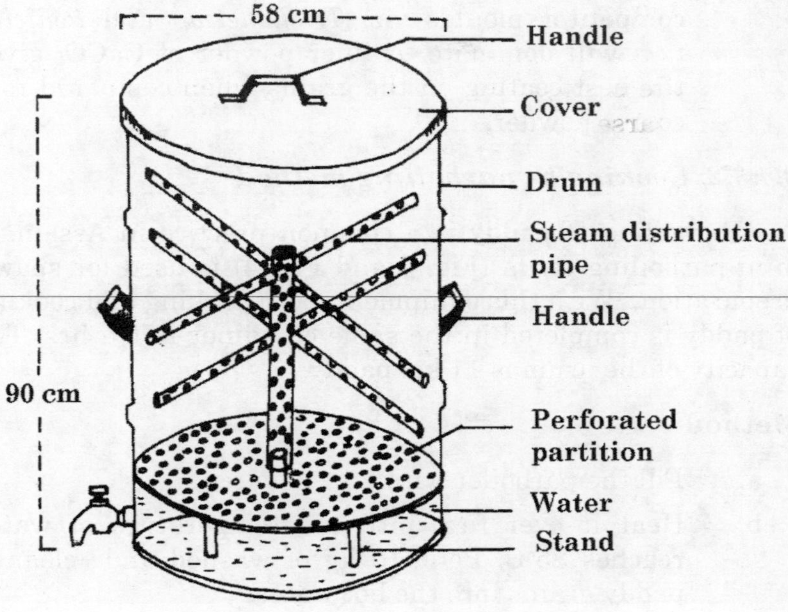

Fig. 4. A mini parboiling drum

c. Reduce the fire and maintain the heat using a piece of firewood.

d. Leave the paddy + water mix as such for 4 hrs. The temperature inside the drum will be stabilized at 65°C.

e. Drain off the water from paddy through the tap. A layer of 2 cm water will be left at the bottom.

f. Resume heating and continue for 30 min. The steam passes through perforated pipes into paddy. A few grains at the top will split, indicating that the process of cooking is complete. There is no need of air drying the grains. The moisture content of cooked paddy will be 36% which is good for growth of mycelia of mushroom. One kg of cooked paddy becomes 1.4 kg after processing.

A plus point with parboiled-paddy spawn is that the chances of bacterial contamination are relatively less. Parboiled paddy is the most suitable substrate for both oyster and paddy straw mushroom in Assam where wheat is not readily available.

3.5. FILLING GRAINS IN BOTTLES AND STERILIZATION FOR MOTHER SPAWN

The mother spawn should be prepared in empty saline bottles. They are convenient for inoculating the spawn packets. Further, the chance of contamination is less in glass bottles.

Method

a. Fill the grains in empty saline bottles up to $3/4^{th}$ capacity i.e. about 300 g/bottle.

b. Put the cotton plug and cover it with a piece of butter paper.

c. Keep the bottles in an autoclave or a pressure cooker for sterilization.

How to use an autoclave for sterilization of grains for spawn?

i. Fill the autoclave with water, up to the level of the stand.

ii. Arrange the bottles one above the other.

iii. Close the lid, tighten the bolts in opposite direction.

iv. Keep open the outlet for steam.

v. Switch on the main.

vi. Allow the air and steam to go out for 10 minutes in order to drive out the air present in the autoclave. If the air is not completely driven out, the desired temperature i.e. 121°C does not develop inside the autoclave. If half of the air inside the autoclave is removed, the temperature will be 112°C at 15lb/ sq. inch. pressure. When all the inner air is removed the temperature will be 121°C at 15lb/ sq.inch. Therefore, never close the outlet without removing the air and steam mixture.

vii. Maintain the pressure at 15lb/sq.inch for 2hrs for spawn bottles and packets.

viii. Switch off the main after sterilization; slightly open the valve for steam outlet. If this is not done, a vacuum is created inside when the pressure falls to zero. As a consequence the inner wall of the autoclave may bulge inside.

ix. Do not open the steam outlet when there is pressure. This will result in blowing off of cotton plugs.

x. Open the lid after complete cooling, take out the materials and transfer them to the inoculation room.

How to use a pressure cooker for sterilization?

i. Pressure cooker of 18 litre or 22 litre capacity can be used.

ii. Place the perforated stand at the bottom.

iii. Fill water up to 10 cm height for 2 hrs sterilization and 4 cm for 30 minutes sterilization.

iv. Place a separator above the water to avoid contact of water with spawn packets or bottles.

v. Fill the materials inside the pressure cooker, and close the lid. Keep open the steam outlet (whistle).

vi. Heat the pressure cooker.

vii. Allow the steam to go out for 5 minutes.

viii. Put the pressure weight.

ix. Note the time when the first whistle is heard. The whistle indicates that the pressure has reached 15 lb/ sq. inch.

x. Reduce the flame to minimum/sim. There should be a slight hissing sound near the steam outlet.

xi. Leave it for 2 hrs for sterilizing substrate for spawn.

xii. Switch off the burner and allow the pressure to come down to zero.

xiii. Open the lid and transfer the materials to inoculation room.

3.6. PREPARING THE INOCULATION CHAMBER FOR WORK

Laminar flow machine is a must for spawn production. The machine is kept inside a small isolated room of 3m x 3m size.

i. Mop the floor of inoculation room with a cloth soaked in water to which 4 spoon full of bleaching powder is added. This is done in the evening hours.

ii. Prepare a solution of formalin @ 10 ml/liter of filtered water. A fine mist of this solution is sprayed in the air, inside the inoculation chamber. The spores of contaminating fungi are suspended in the air inside the inoculation chamber. When a fine mist of formalin solution is sprayed in to air, these suspended spores will be brought down along with the mist particles. This operation is done in the evening, prior to the day of inoculation.

iii. Arrange all the materials viz. spirit lamp, match box, inoculation needle, cotton and the substrate packet for spawn on the working space of the laminar flow. Always use wax coated match sticks to light the spirit lamp. Mould will grow on ordinary match sticks due to humidity, which is a source of contamination.

iv. Switch on the UV light for 20 minutes just before inoculation.

v. Wash your hands with antiseptic soap before going inside the inoculation room.

3.7. INOCULATION OF SUBSTRATE FOR MOTHER SPAWN

The second generation (T_2) fungal culture prepared from mother culture (T_1) on the sterilized grain is called mother spawn (Fig. 5).

Method

a. Keep the sterilized bottles containing grains in the inoculation room.

b. Keep ready the spirit lamp, cotton, wax coated matchstick box and inoculation needle on the working table of laminar flow cabinet. Switch on the UV light for 20 minutes.

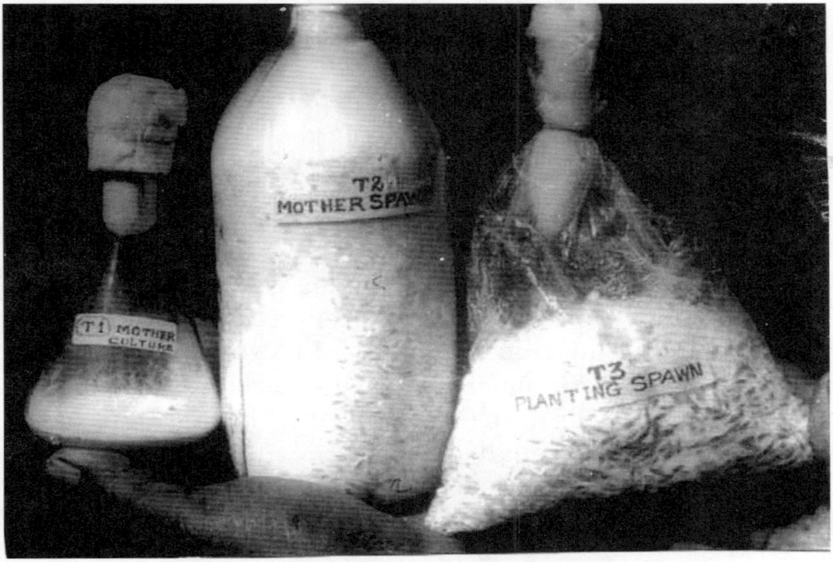

Fig. 5. Three stages of the culture of mushroom fungus: Mother culture (T1), Mother spawn (T2) and Planting Spawn (T3)

c. Cut a piece of about 1 cm x 2 cm size agar along with growing mycelium with the help of a stiff inoculating needle. Lift the piece with the hook of the inoculating needle. Transfer the mycelium aseptically into the mother spawn bottle. Two pieces of mycelium may be put in a bottle. The spawn growth is fast when the size of inoculum is big. Put a sticker label with the name of species, T_2 mark and the date. Five spawn bottles can be inoculated from a T_1 culture prepared in 100ml conical flask.

d. Keep the inoculated bottles in an incubator at 24°C for *Pleurotus* spp. and *Agaricus bisporus*. *Volvariella* sp. and *Calocybe indica* should be kept at 28 to 30°C.

e. Watch the growth of mother spawn frequently. Discard immediately the contaminated bottles.

f. The growth of the mother spawn will be completed after 15 days of inoculation. Allow the spawn to mature for 5 more days. The growth of the mycelium will become dense. The spawn is now ready for putting into planting spawn packets.

3.8. PREPARATION OF PLANTING SPAWN OR COMMERCIAL SPAWN

Commercial spawn (Fig. 5) is prepared in polypropylene bags by multiplying from the mother spawn up to third (T_3) or fourth (T_4) generation. This spawn is used for planting in the mushroom beds. Further sub culturing of the spawn from T_4 generation usually reduces the efficiency and vigor of the spawn which leads to reduction in mushroom production.

Method (Fig. 6)

a. Take polypropylene bags of 15 x 21 cm size. The bags should have double sealing at the bottom, or else water may enter during autoclaving.

b. Fill 200g of the substrate (cooked paddy grain).

c. Put a cotton plug for aeration into the spawn and tie it with a rubber band. Cotton plug is a must for aeration and respiration of growing mycelium.

d. Sterilize them at 15 lbs/sq. inch for 2 hrs.

e. Keep the sterilized bags on the laminar flow under UV light for 30 min.

f. Select a fully grown mother spawn (T_2) bottle.

g. Involve two persons in transferring the inoculum from the mother spawn bottle to polypropylene bags.

h. One person will hold the bottle horizontally in left hand, open the plug and hold it in between the 4th and last finger of the right hand.

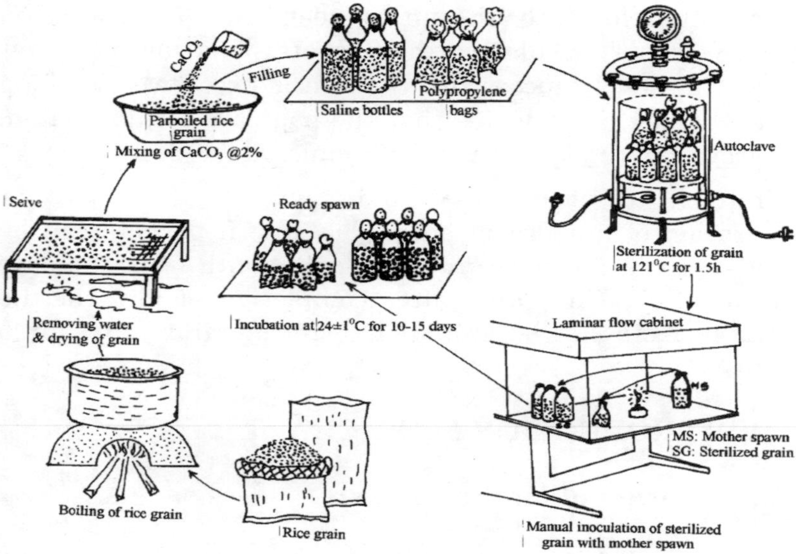

Fig. 6. Preparation of mushroom spawn

i. Second person will slightly open the mouth of the polypropylene bag.

j. The first person will transfer two spoon full of mother spawn to the polypropylene packet. With one bottle of mother spawn, 50 spawn packets of 200g each can be inoculated.

k. The mouth of the packet is closed quickly with a cotton plug, to minimize the risk of contamination. Cotton plug is a must for aeration and respiration of growing mycelium.

l. A label sticker is put on the packet showing the species, generation and date.

m. Incubate the spawn packets at the temperatures specified earlier for different species. The chances of spoilage due to *Trichoderma* sp. is more at warm temperature (above 25^{0}C). The best temperature for *Pleurotus* sp. is is 23-24^{0}C.

A cotton like growth can be seen from the inoculum after 48 hrs. of inoculating the packets. This confirms that the inoculation is successful. If there is excessive sweating inside the bag, it indicates that the bags are contaminated. The growth of the spawn will be completed in 12 to 15 days.

Fifteen to twenty days old spawn is the best for preparation of mushroom beds. The spawn packets can be stored at room temperature up to one month from the date of inoculation. Generally, after completion of growth, the spawn packets can be stored in a refrigerator (4°C) for 3 months.

SPAWN PRODUCTION FLOW CHART

A. Preparation of Mother Spawn

Healthy, washed paddy or wheat grain

↓

Boil grains in water for 30 min

↓

Drain out water on a sieve

↓

Dry the grain under shade for 1 h

↓

Mix calcium carbonate @ 20 g/kg cooked grain

↓

Fill 300 g grain in empty saline bottles

↓

Put cotton plug and autoclave at 15 lbs/sq. inch for 1.5 to 2 hrs.

↓

Inoculate the bottles with a piece of growing mycelium

↓

Incubate in B.O.D. incubator at 23-24°C for 12-15 days

↓

Mother spawn is ready

B. Preparation of planting spawn (T3)

Cook the grains as stated

↓

Mix calcium carbonate @ 20 g/kg

↓

Fill polypropylene bags with grains

↓

Autoclave at 15 lbs/sq. inch for 1.5-2 hrs

↓

Inoculate 2 spoonfull of grain from mother spawn to each bag

↓

Incubate at 23-24°C for 12-15 days

↓

Planting spawn is ready

3.9. PROBLEMS FACED DURING SPAWN PRODUCTION

3.9.1. Slow growth of spawn

Sometimes it takes about one month for completing the growth of the spawn. The cause of slow growth is the use of cheaper grade of Calcium carbonate ($CaCo_3$) which is known as French chalk, sold in hardware shops at about Rs.6/-kg. The pH of the grain substrate is the crucial deciding factor for healthy growth of the spawn. Slow growth of spawn is the consequence of below optimum level of pH (above 6.5) of the substrate. The pH becomes less than 6.5 when French chalk is used. Hence $CaCO_3$ of pharmaceutical grade should be used to overcome this problem.

3.9.2. Bacterial contamination

Spawn is spoiled sometimes due to bacterial contamination. Symptom of this contamination is the accumulation of a slimy fluid in between the grains and the inner wall of the spawn bag. Consequently the advancing mycelial growth is ceased or stopped or checked by the bacteria. An inhibition zone resembling island is seen in the spawn bag. Wheat grain is more prone to bacterial contamination than paddy grain. The excess moisture content in wheat is responsible for favoring bacterial contamination. *Bacillus brevis* is the species generally found contaminating the mushroom spawn.

3.9.3. Spoilage of spawn by moulds

The fungi responsible for spoilage of spawn are *Trichoderma* sp., *Aspergillus* sp., *Penicillium* sp. and *Rhizopus stolonifer*. The following factors contribute to the growth of these fungi,

a. Contamination is high when paddy grains are not properly sterilized i.e. sterilized for less than 2 h duration.

b. Chances of spoilage are high when the atmospheric temperature is high (above 27°C).

c. Contamination is more when bad quality paddy grain is used, i.e. grains damaged by insects.

Management

i. The best duration for sterilization of grain is 2 h at 15 lbs/sq. inch in order to reduce the chances of contamination.

ii. Add Thiophanate methyl or Carbendazim @ 0.5g/kg to cooked grain before autoclaving for checking *Trichoderma* and *Penicillium* spp.

3.10. STORAGE OF SPAWN

(i) Fifteen to twenty days old spawn is the best for preparing mushroom beds. (ii) Spawn can be stored under room temperature up to 30 days from the date of inoculation. (iii) After one month the spawn can be stored in a refrigerator for another 3 months. However, there will be some decrease in yield of mushroom with increase in storage period above 2 months.

CHAPTER IV

Cultivation of Oyster Mushroom

The fruit bodies of this mushroom resemble the shell of sea oyster and hence called oyster mushroom (*Pleurotus* spp., Fig. 7). This mushroom is widely cultivated in Tamil Nadu, Kerala, Andhra Pradesh, West Bengal and Assam than any other parts of India. Advantages of oyster mushroom over button mushroom or paddy straw mushroom are as follows:

i. Cultivation technology is simple and easy to adapt

ii. Not exigent in its temperature requirement – 20° to 30°C

iii. Possesses the highest bioconversion ability – >60%

iv. Its growing season is longer – 8 months in a year

v. Can be easily sun dried.

4.1. GROWING SEASON

The best temperature for oyster mushroom is 23° to 25° C. The best months of the year for growing this mushroom in Assam is October to April.

Fig. 7. Oyxter mushroom

4.2. SUBSTRATE

Rice straw is the best substrate among all the cellulosic materials. The straw should be of good quality – golden yellow in colour. If the straw is exposed to rain, it can't be used. There will be a lot of contamination of beds when straw from *ahu* crop (harvested in July) is used.

4.3. SPECIES

There are seven species of oyster mushroom grown in Assam. These are *Pleurotus sajor-caju* (grey oyster mushroom), *P. sapidus, P. citrinopileatus, P. cornucopie, P. florida, P. flabellatus* and *P. ostreatus* (Hirakitake). The most beautiful oyster mushroom is *P. eous* (Pink mushroom) for its attractive colour. It was not brought to the commercial cultivation because of its less yield and slightly leathery in texture.

4.4. CULTIVATION METHOD

4.4.(A): Polythene bag method

4.4.1. Making the substrate ready for spawning

i. Cut the straw into pieces of 5 cm long. This can be done by flywheel type of chaff cutter with only one blade. The straw pieces will be too small when two blades are fitted to the flywheel. Alternately, the straw can be chopped with a hachette.

ii. Soak the straw for 6 to 8 hrs in an empty oil drum. Place a heavy weight to prevent floating of straw.

iii. Drain out the water.

iv. Sterilization of substrate in boiling water: Water is boiled in an aluminum saucepan or in a G.I. drum to 85°C. Keep the wet straw immersed in boiling water for 30 minutes. The lid should be closed and there should be slight heating to maintain the temperature. The success of cultivation depends on boiling. The aim of this process is to eliminate unwanted micro-organisms from the straw. Besides the mushroom fungus breaks down cellulose in the straw quickly and establishes itself fast in boiled straw.

v. Remove the straw from boiling water. The hot water should be reused for boiling a second lot of straw. One kg of straw becomes three kg after soaking and boiling.

vi. Spread the straw on a clean floor for cooling. There should not be excess water in straw. When straw is squeezed in palm, water should not drip down.

vii. Spray Carbendazim @0.5 g/l. Two hundred ml is sufficient to spray the straw required for one bed. As an alternative to Carbendazim, 0.3% garlic extract can be sprayed on the straw just prior to spawning. This is very effective against green mould (*Trichoderma* sp.).

How to prepare garlic extract?

a. Take 50 g of peeled garlic cloves and make a paste.

b. Add 50 ml of water to the paste. Squeeze it through muslin cloth to get the stock solution. Take 3 ml of stock solution and add to 1 liter of water. It becomes 0.3% garlic extract which can be sprayed on straw. Garlic extract prevents contamination of straw beds due to green moulds.

Chemical sterilization of substrate

This method is an alternative to the above mentioned hot water sterilization method (steps iv. and v).

The principle: Straw is treated by soaking in a chemical solutions of carbendazim 50% WP (75 ppm) and formaldehyde (500 ppm) for a period of 16-18 h. Most of the competitor moulds are killed or their growth is suppressed for 25-40 days after spawning.

This technique was developed by Vijay and Sohi from NRCM, Solan (H.P.) in 1987. The technique is as follows: Ninety litres of water is taken in a rust proof drum of 200 l of water capacity. Twelve kg of straw is slowly steeped in water. In another plastic bucket 7.5g carbendazim and 125 ml of formaldehyde (37-40%) are dissolved in 10 l of water and slowly poured on the already soaked straw. Straw is pressed and covered with a polythene sheet. After 15-18 h the straw is taken out and excess water is drained.

4.4.2. Making ready the polythene bag

a. Take a polythene bag of 40 cm x 60 cm size. Fold it lengthwise, 4 times. High density opaque polythene bags are better than transparent polythene bags for spawn running.

b. Make 20 holes with the help of single punch machine.

c. Tie the bottom of the bag with a jute thread. This gives a circular bottom to the bed.

4.4.3. Making ready the spawn

a. The rate of spawn per 3 kg wet straw (1 kg initial dry weight) is 3% i.e. 30 g per kg. Usually 100 g spawn is required for one bed of mushroom.

b. Break the lumps of planting spawn with fingers. Divide the 200 g spawn into two parts of 100g each in a paper plate. Make four equal parts out of the 100g.

4.4.4. Making ready ventilation pipe

Take a PVC pipe of 3 cm diameter and 35 cm in length. Make holes at 5 cm interval on two opposite sides.

4.4.5. Spawning

a. Put the layer of 10 cm straw in the bottom of the polythene bag. Press it with palms to let the air go out and the layer becomes 6 cm in height. The ventilation pipe is than put at the center.

b. Twenty five grams of spawn is sprinkled over the straw putting more towards the sides.

c. The bag is filled with five layers of straw and four layers of spawn. The bag is than tied with a jute thread along with a label of the species and date of spawning.

4.4.6. Spawn running

The mushroom beds are arranged in a spawn running room. It is thatched shed of 3m x 6m. Bamboo racks are built to accommodate maximum number of beds. This room does not require light. The mycelial run is faster in a dark place as compared to light. The best temperature for spawn run is 25°C and it is completed in 14 days. Additional 4 days are needed for the maturity of the mycelium. Cream coloured patches on the surface which indicate the maturity of beds.

4.4.7. Opening of mushroom beds

Open the bag; place the left hand with the ventilation pipe in between the thumb and index finger. Hold the bottom end with the right hand and invert the bag. With slight jerks, the bed is taken out from the polythene bag.

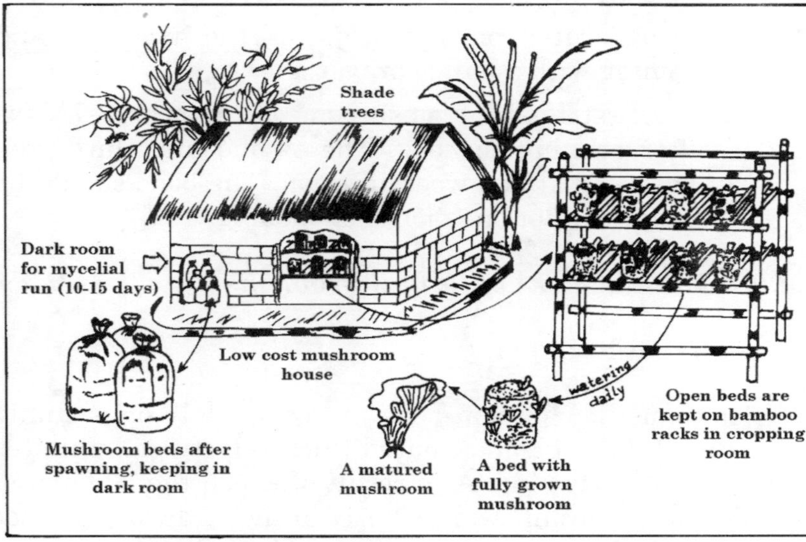

Fig. 8. Cultivation of Oyster Mushroom

4.4.8. Cropping room

The beds after opening are kept inside the cropping room of 3 m x 6 m size with thatched roof. There should be diffused light and ventilation. The best temperature is 23°C.

A 5 cm layer of sand is put on the floor and the walls of the room are lined with hacianth cloth on the inner sides. The floor and the walls are watered, usually twice a day to maintain sufficient humidity. The principle is that, the rate of evaporation from the mushroom beds should be reduced. Direct sunlight should not fall inside the cropping room. The best place of a cropping house is under the shade of a big tree (Fig. 8).

4.4.9. Care during Cropping Period

i. Transfer the beds to cropping room.

ii. Water need not be sprayed on the beds for first two days after removing the polythene.

iii. Spray water on the beds in the morning and evening with a hand sprayer.

iv. The first flush of mushroom will appear 6-7 days after opening the bag. The second and third crop will come at one week interval. The beds should be removed after one month of opening.

4.4.9.1. Management of fungal problems

a. Green mould in beds

Sometimes mushroom beds are completely or partially covered with light green coloured thick velvety growth after 7 to 8 days of spawning. As a result of which the mushroom mycelia can not proliferate properly in the straw or does not grow at all during spawn running period. The fungus, *Trichoderma harzianum* is responsible for this type of spoilage of mushroom beds (Fig. 9). Usually this is a common problem occurring during May to September as the high temperature and high relative humidity of this period are very congenial for *T. harzianum*. There are two causes for this problem: (i) hot water treatment of straw is not proper, (ii) Contaminated spawn.

Fig. 9. A bed of oyster mushroom spoiled by green mould.

Management: (i) Spray Carbendazim @0.5g/l or Garlic extract @0.3% on pasteurized straw before spawning as mentioned earlier, (ii) Break the spawn packet and check carefully for any fungal contamination before spawning. Discard the contaminated spawn. Do not throw them near the mushroom house. (iii) Wash the used mushroom bag very cleanly immediately after removal from the beds and air dry properly before reuse. (iv) Remove all the contaminated beds from the mushroom house and burry them to check its spread to the healthy beds.

b. Ink cap fungus

It is a weed mushroom. The fungus *Coprinus* spp. is known as ink cap. It has a 5 –7 cm long and hollow stipe with a gray coloured umbrella like thin pileus at the top. The lanky fungus grows singly or in groups. The affected beds turn into black to deep blue colour due to severe dropping of the spores of this fungus. As a result fruiting of mushroom becomes erratic and sporadic or even does not occur. The cause is due to the use of bad or decomposed straw due to rain.

Management: (i) Use good and properly dried straw for bed preparation. (ii) Use the straw of *Sali* rice (harvested in November- December). Do not use the straw of *Ahu* rice (harvested in June-July). (iii) Remove all the ink caps at young stage whenever seen on the beds.

4.4.9.2. Management of Bacterial problem

a. Browning: Yellow to light brown colouration develops on the margin of the oyster mushroom and its stalk. Due to the infection or growth of a bacterium *Pseudomonas* sp., browning of the fruit bodies takes place. Main cause of browning is the accumulation and deposition of water on the fruit bodies after watering on the beds.

Management: (a) Avoid direct watering on the mushrooms, (b) Do not let deposit water droplets on the mushroom fruit bodies. Hence slightly shake the beds after watering and (c) Spray Bleaching powder @ 2g per 10 l of water twice a week after harvesting.

4.4.9.3 Insect problems

There are three types of flies that infest oyster mushroom during summer season. The most favourable temperature for their prevalence is 27-30°CF and their life cycle completes within 2-4 weeks.

a. Mushroom fly

i. Sciarid fly: The adult of this fly is slender, body dark brown, long back with long thin legs, long delicate wings and has thread like antennae. Maggots (Dipteran larvae) are legless, white, 6-7 mm long, head black with powerful chewing type of mouthparts.

Nature of Damage: Adults lay eggs in the gills and the maggots emerging out from eggs. The adult flies do no cause damage to the mushroom fruit bodies. Maggots hiding in the gills eat away the soft tissue of the gills causing distortion of the fruit bodies. Moreover they chew up mushroom, make tunnels in the pileus and stipes and as a result the mushroom becomes unfit for consumption.

Fig. 10. Phorid fly

ii. Phorid fly: Adults resemble to those fruit flies (Fig. 10). They are shorter and thicker than the Sciarid flies, possess smaller head, larger chest and 'hump back' and very active with jerky run. Maggots are legless, white, 4 mm long, have slightly pointed head and tail (Fig. 11).

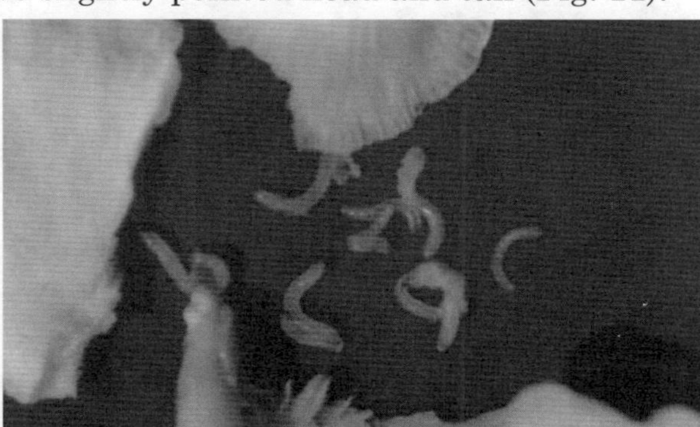

Fig. 11. Larvae of phorid fly

Damage: As Sciarid fly.

iii. Cecid fly: Adults have delicate midges. Maggots are orange or cream coloured, 2.2 mm long and reproduce without becoming adult (paedogenesis).

Nature of Damage: Grubs feed on the mycelia running on the substrate apart from damaging the soft fruit bodies.

b. Beetles

i. Staphylinid beetle: (Fig. 12)

Adults: Dark brown in colour, very active with short elytra and large membranous hind legs. The tip of the abdomen is held curled over the back. The adults do not cause damage to mushrooms. The adult beetles are predatory on smaller organisms like mites and collembolans.

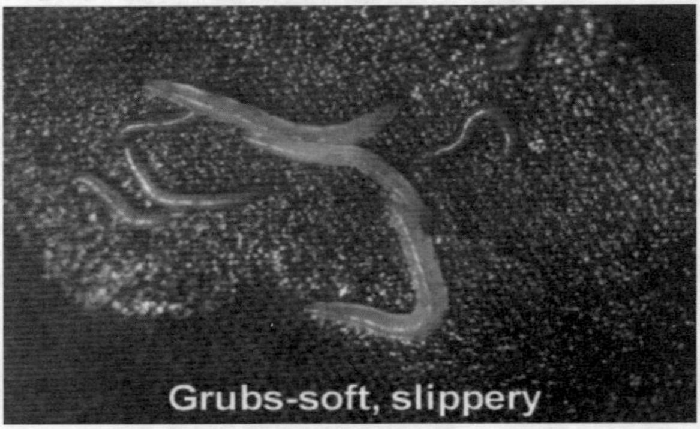

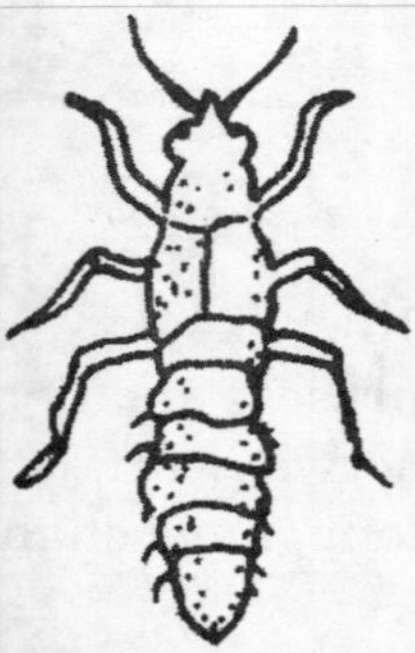

Adult beetle

Fig. 12. Staphylinid beetle - A pest of Oyster mushroom

Grubs: Whitish, long cylindrical with tubular terminal segments. Their body is very soft and slippery. The grubs hide between the gills. They make small irregular holes in the hymenium and the stipe. The infested mushroom turn brownish and decay. The grubs damage the over mature nad decaying mushrooms.

Predisposing factors: The smell of decayed mushrooms attract the adults. They lay eggs on the mature or discarded mushrooms. The emerging grubs feed on the soft gills and crawl over the beds. This pest occur during May to August.

ii. Scaphisoma beetles (Scaphidiidae: Coleoptera): Two species viz. *Scaphisoma tetrastictum* and *S. nigrafasciatum* of Scaphidiidae, Coleoptera cause damage to the fresh fruit bodies.

Scaphisoma tetrastictum: Adult beetles are deep amber coloured, oval shaped, length about 1.95 mm, breadth 0.82 mm through mesothorax, head hypognathus and tip of the abdomen is not fully covered by elytra. Grubs are dirty white, elongate (0.6- 1.2 cm), campodiform, antennae short, thorasic legs well developed and body glabrous. Pupa are creamy white and 2-3 mm long.

S. nigrafasciatum: The beetle is smaller than *S. tetrastictum*. Adult beetles are shining black, elongate oval, length 1.55 mm, breadth 0.60 mm through mesothorax, head hypognathus, tip of the abdomen is protruding below elytra and antennae long filiform.

Predisposing factors: (a) The smell of decayed mushrooms attract the adults to lay eggs on mature cap or pileus and (b) Discarded debris.

Nature of damage: Adults feed the soft tissues of stipes, gills and pileus and also feed mycelia during mycelial running. The grubs make irregular holes in the hymeniun and stipes. They hide between the gills and the base of the stalks. The growth of the fruit bodies becomes very poor due to infestation by these beetles. Infestation by the grubs reaches the peak during June to August.

iii. Pleasant beetle (Erotylidae: Coleoptera): The beetle belongs to the family Erotylidae of the order Coleoptera. Fresh specimen of the adults are shining orange in colour that turns into dark brown while preserved in 70% alcohol (Fig. 13). They are oval and convex measuring 4 mm x 3 mm. Grubs are elongate, 8- 9 mm long, white, head dark brown.

Predisposing factors: As (ii).

Nature of damage: Grubs are more aggressive than the adults. They damage the fruit bodies by eating the soft tissues and making number of holes. The grubs are present in between the gills.

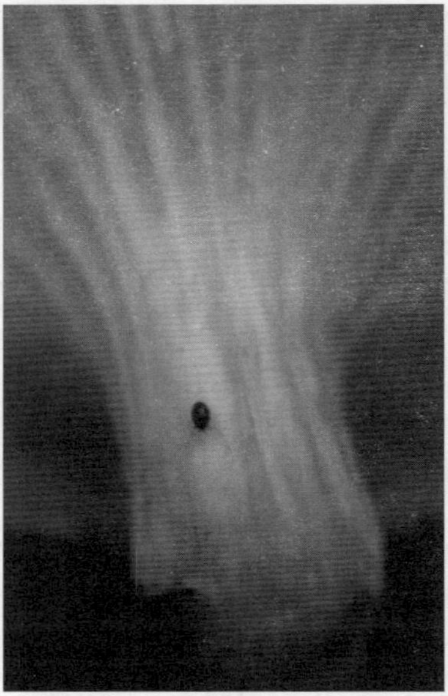

Fig. 13. Pleasant beetle

c. Others

Slugs and snails also cause more or less spoilage of the fresh mushrooms by feeding the soft tissues. Generally they crawl over the mushroom beds and secrete many slimy and slippery threads giving an obnoxious look to the beds and fruit bodies.

Management: (i) Remove over matured mushrooms at the earliest. This prevents the adults from laying of eggs. (ii) Apply bleaching powder in the mushroom house and its premises. Bleaching powder has a repellent effect on the beetles. (iii) Spray neem-based insecticides (Azadirachtin 0.15 EC) like Rakshak, Neemagon, Neemajol @ 3ml per litre of water before spawning of the boiled straw and after opening of the beds. One more spray should be given on the beds after the first harvest. (iv) Spray Endosulfan @ 1ml/l on straw at the time of spawning. This is superior to neem based products. (v) Spray chemical insecticides like Malathion 50 EC or Nuvan @ 1 ml per litre immediately after opening of the beds. Malathion is not systemic. It remains on the surface and therefore it can be washed off. Yet, never spray these insecticides directly on the growing fruit bodies. (vi) Fix nylon net (mosquito net) on the ventilations and windows of the mushroom house to prevent entry of the flies and beetles. (vii) Smoke daily or at alternate days to get rid of the insects from the mushroom house. (viii) Light traps: Polythene sheet is smeared with a sticky material viz., mustard oil and attached to a yellow colour bulb in cropping room help in controlling adult flies. Insects are attracted to white light at temperature above 15^0C and to yellow light at lower temperature.

4.4.(B) CUBE CULTURE METHOD

The steps, up to boiling of straw are the same as in case of polythene bag culture of oyster mushroom. Other steps are as follows,

- Remove the straw from the boiling water and allow to cool. Squeeze out the excess water.
- Punch holes at 6 inch intervals in a polythene sheet of 1m X 1m size. Spread the sheet on a table on which a binding thread has already been placed.
- Place the mould (Fig. 14) in the center of the polythene sheet.

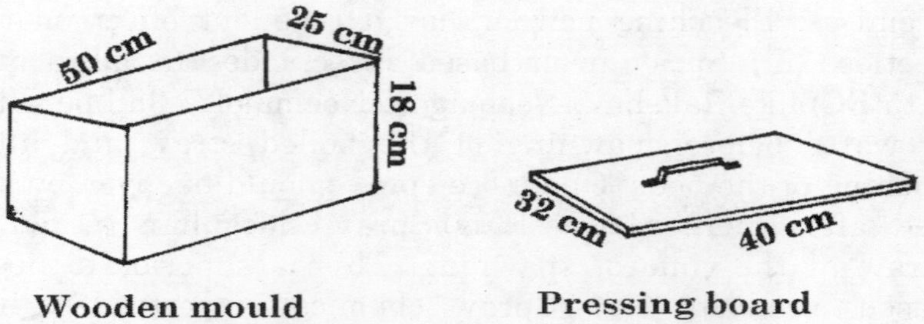

Fig. 14. A mould for cube culture

- Break the spawn into small bits. Two hundred grams of grain spawn is required for one cube.
- Put about 1 kg of straw and spread it in mould uniformly, then slightly press it with hand. Some spawn bits are uniformly distributed over the straw to complete the first layer of spawning.
- Put another 1 kg of straw and make it uniform. Press it hard with pressing board and distribute the spawn bits uniformly over the second layer.
- Repeat the process six times ensuring that there is proper pressing for every two layers. Two hundred and fifty gram of spawn is required for making one cube.
- Remove the mould by pulling upwards and pressing the straw downwards.
- Wrap the cube with polythene sheet and bind it with thread. Keep the cube in spawn running room for 18 days.
- Remove the polythene sheet and keep the cube in the cropping room (Fig. 15).

4. 5. HARVESTING

It is possible to obtain complete food value, taste and medicinal values by consuming mushroom in fresh condition that is harvested at the right stage. The mushrooms harvested lately after attaining full maturity can not provide the wholesome food values. Mushrooms should be used either freshly for consumption or to be preserved for future use. The best mushrooms will be slippery when touched.

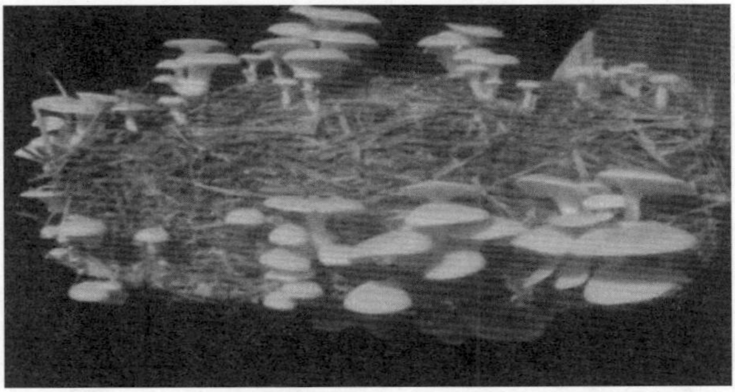

Fig. 15. Oyster mushroom grown by cube culture method

When and how to harvest ?

Young cluster of fruit bodies will turn into fully developed mushroom after 3 to 4 days of their appearance. Then the margin of some fruit bodies (pileus) starts curling upward or downward. This is the right stage for harvesting. This fruit bodies should be harvested by uprooting from their bases. They should not be plucked. Otherwise the bases will be left on the bed, rot later and cause spoilage of the mushroom beds.

How to do packing ?

i. After harvesting, the stalks are trimmed to remove any adhering straw.

ii. Mushrooms are packed in brown paper bags with 250g contents. They can also be packed in perforated polypropylene bags.

4.6 PRESERVATION OF MUSHROOM

Mushroom has a very short shelf life due to its highly perishable nature. Therefore it is not possible preserve for many days in fresh condition. However it can be stored for desired time period adopting some procedures that have been mentioned below.

4.6.1 Reasons of quick rotting

1. Fresh mushrooms possess more than 90% water. This water content is lost through evaporation and respiration after harvesting. As a result quick withering of the fruit bodies takes place.
2. Many physiological and metabolic activities in the fruit bodies do not stop even after harvesting that lead to deterioration of the mushrooms.
3. Higher water content of the fresh mushroom invites many microbes, airborne fungi and bacteria that multiply in the tissue of the harvested mushroom which finally cause rotting.

4.6.2 Selection of mushroom prior to preservation

i. All the mushrooms are not equally suitable to store. For example, too young and over matured mushrooms are can not fetch good price in the market. The right stage of harvesting of mushroom for preservation is

   Button mushroom : Button stage

   Paddy straw mushroom : Button stage

   Oyster mushroom : Margin of the pileus when starts to curl

ii. Mushrooms of the first flush (in the case of oyster) are the best for preservation. Because, the colour and taste of this batch remain intact almost in the natural form during the time of preservation.

iii. Spotted, deformed, insect damaged fruit bodies are to be discarded.

iv. Fruit bodies should be sorted out based on the size and age to maintain uniformity.

4.6.3 Types and methods of preservation

Many methods have been developed in order to keep the mushroom for a desired period. These methods have been grouped into two according to the duration of preservation.

1. Short term preservation.
2. Long term preservation.

4.6.3.1 Short term preservation

This means preservation for maximum of 10-15 days only. The methods adopted for this type of preservation are simple and the required materials are very easily available. The main aim of preservation for short period is to increase the shelf life of the mushroom by,

i. Suppressing microbial attack on the mushroom.

ii. Reducing and regulating the physiological activities in the fruit bodies.

iii. Reducing the rate of water loss or evaporation from the harvest.

iv. By keeping in different chemical solutions.

Methods

i. Keeping in room temperature: Mushrooms can be kept in the room temperature by putting in the paper bags. During summer days at 30-33^{0}C, the mushrooms remain fresh for 8-12 h only without deterioration of the quality whereas it is possible for 24-36 h during winter days.

ii. Storage in refrigerator: Fresh mushrooms can be stored for 7-15 days in a refrigerator. However this duration varies according to the storage temperature. e.g.

Sl. No.	Type of package	Storage temp. (^0C)	Duration of Storability (days)
		0	13
1	100 gauze polypropylene bag	4	9
		7	6
2	Non-porus polypropylene bag	5	4
3	Porus polypropylene bag	10-15	4

iii. Preservation in brine solution: Water with common salt at high concentration (10-15%) is suitable for preservation of fresh mushroom for 6-7 days. The solution should be stirred occasionally and extra salt is to be put to maintain the concentration.

iv. Other methods

a. Lactic acid fermentation : Tedious method.

b. Gamma irradiation : Costly and risky method.

4.6.3.2 Long term preservation

Both economic and expensive processing methods are applied for this purpose. In some methods, the shape, quality and taste of the mushrooms are reduced whereas in some other methods these qualities are enhanced. After processing, mushrooms are packed and canned in the airtight containers. Prolonged preservation with good quality is possible through proper processing and canning.

Methods

i. Sun drying: The principle of this method is dehydration. This is the most easiest and economic method.

After sorting and selection, fresh mushrooms are thinly spread on a sieve and sun dried for 2-3 days. To avoid browning of the fruit bodies due to direct sun light, a shade may be provided to the mushrooms by spreading a black cloth at about 1 feet above the sieve. Since the base of the stipe is hard in texture, not palatable like the pileus and also can not dried up properly, hence this part of the mushroom should be trimmed during sorting. In this process, the weight of the mushroom is lost by 1/10-1/12 of the fresh weight. Then the dry mushroom should be packed in small lots each weighing 25-50 g in 100 gauze polypropylene bags and sealed or in airtight containers and stored in dry, cool place. Thus mushroom can be preserved up to 5-6 months for table purpose.

ii. Machine drying: Its principle is same with the sun drying method Fresh mushrooms are dried in the electrically operated drier within 6-8 hours. Advantage of using a drier is that mushrooms can be dried even during cloudy and rainy days. This is a costly method of drying and has to depend upon constant power supply.

iii. Blanching: In this method, fresh mushrooms are boiled for a given time period and then sun dried. Small quantity of the fresh mushroom is held in a muslin cloth after sorting into different groups according to their sizes and steeped in warm water of 80-85^{0}C for 1.5-2 min only. Then the mushrooms should be taken out of the warm water, spread very thinly on a sieve and dried in the sun or in a drier. To retain or improve the natural colour of the mushroom, some chemicals viz. potassium metabisulphate (KMS) @300 ppm, sodium chloride (NaCl) @400 ppm and citric acid @0.1-0.2% may be put into the water before boiling. These chemicals can be used singly and also in combination. The mushrooms such processed can be shelved for long time. They are not easily spoiled by moulds. It is worth mentioning that mushrooms blanched in KMS solution are not good for the heart diseased patients.

iv. Other methods

a. Freeze drying

b. Canning

c. Steeping in chemical solution

d. Pickling.

CHAPTER V

Cultivation of Paddy Straw Mushroom

Paddy straw mushroom is called tropical mushroom as this mushroom likes warm humid weather which is prevalent during May to September in Assam.

5.1. ADVANTAGES

i. It is the quickest growing species among mushrooms. It takes only 14 days from sowing to harvesting.

ii. It tastes better as compared to oyster.

iii. Method of cultivation is easy – hot water treatment is not essential, unlike for oyster.

iv. It needs only simple equipment – cement tank, bamboo platform and straw.

5.2. DISADVANTAGES

i. Keeping quality of this mushroom is very poor. It should be consumed on the day of picking. When stored beyond 24 hrs. mushrooms become watery, emitting smell.

ii. Huge quantity of straw is required; 100kg of straw (dry weight) gives only 13kg of mushrooms. The biological efficiency is 13-15%. In contrast, 70kg of oyster mushrooms are harvested from 100kg of straw.

5.3. SPECIES

There are three species namely *Volvariella volvacea*, *V. esculenta* and *V. diplasia*. Of these, *V. volvacea* is widely grown (Fig. 16).

Fig. 16. Paddy straw mushroom

5.4. GROWING SEASON

Volvariella volvacea grows well between 28 to 35°C. This range of temperature is available during the months of May, June, July, August and September in Assam. The month of August supports maximum production. The optimum range of temperature for *V. volvacea* is as follows,

Growth stage	Optimum temperature	Temperature range
Mycelial growth	32° to 35°C	15° to 45°C
Fruiting	28° to 32°C	22° to 38°C

5.5. BEST pH FOR GROWTH

Paddy straw mushroom prefers higher pH ranging from 6.8 to 7.8.

5.6. METHOD OF CULTIVATION

There are three methods, viz. (i) Standard bed method, (ii) Hollow bed method and (iii) Cage method.

5.6.1 Standard bed method (Fig. 17)

i. Tieing of bundles: This is a special feature for this mushroom. Straw is tied into bundles of 85cm in length and 10 cm in diameter. Each bundle weighed about 800g. Twenty four such bundles are required to construct one bed of mushroom. Eighteen kg of straw is used in one bed. Smaller beds of 10 kg straw are also suitable.

ii. Trimming of the straw bundles: Both the ends of the bundles are chopped with a hatchet to give sharp ends. This step is necessary as the mushrooms arise mostly from the cut ends of the bundles.

iii. Soaking of straw: Bundles are put in an empty oil drum or in a cement tank (1 m diameter,1 m height). A boulder is placed above the straw to stop floating. Water is filled to a level above the straw. The bundles should remain immersed for minimum 8 h and maximum 16 h, after that straw is taken out. Excess water is drained off by holding the bundles for 30 min. on bamboo rack. Like oyster mushroom hot water treatment of the straw is not required for this mushroom.

iv. Cropping room: It is 3 m in breadth, 6m in length and 1.5 m in height at the sides. The roof is of thatch and the floor should be cemented. A 5cm layer of sand is spread on the floor. A lining of plastic sheet is given on the inner side of walls and the roof in order to retain humidity and maintain the temperature. This structure can hold 8 beds at a time. The sand should be soaked with plenty of water.

v. Quality of spawn: Paddy grain is the best base substrate for making the spawn. It should be prepared from the first generation tissue culture (T_1) of mushroom. The biggest problem with *Volvariella* is its fast degeneration during sub culturing. The culture becomes sterile with cottony growth. Such culture produces a few or no mushrooms. The identifying mark of vigorous and fertile spawn is the presence of a layer of brick red stroma on the inner wall of spawn bags/bottles (Fig. 18). Therefore, the deciding factor for a successful crop is the quality of spawn. It should be of T_3 generation with brick red stroma. During off season, the cultures of *Volvariella* should be incubated at 28°C. It should be sub cultured at 3 months interval. Never keep *Volvariella* cultures in a refrigerator. It dies when kept in a refrigerator.

vi. Pigeon pea (arhar dal) powder as supplement: Pigeon pea dal powder is absolutely necessary for quick growth of mycelium and optimum yield. One hundred and twenty grams of dal powder are required for a standard bed. It is divided into 3 parts, viz. 30g, 30g and 60g for applying in 1st, 2nd and 3rd layer of bed.

vii. Steps in construction of mushroom bed:

   a. Place 6 bundles of straw closely in a row on a bamboo platform of 90cm x 90cm x 10cm height. The purpose of the platform is to avoid direct contact of beds with floor and to facilitate draining off excess water.

b. Spray Ethion @ 2.5ml/l on all the straw bundles to prevent straw mites.

c. Dust 25g chalk powder (impure grade) to create the optimum pH. The best pH for the growth of this mushroom is 7.

d. Place small quantities (2g) of spawn in rows at 5cm distance over the bundles of straw leaving a margin of 7cm on all sides of bed without spawning. Fourty grams of spawn are spread on this layer.

e. Sprinkle 30g of arhar dal powder over the bits of spawn.

f. Put the 2nd layer of 6 bundles in the opposite direction of the 1st layer.

g. Sprinkle 25g of chalk powder, sow 40g of seed and sprinkle 30g of dal powder as it was done on the 1st layer.

h. Put the 3rd layer of 6 bundles in the opposite direction of the 2nd layer.

i. Sprinkle 50g of chalk powder, sow the seed (80g) and sprinkle dal powder (60g) over the seed. The seed rate for the 3rd layer is double of that in 1st or 2nd layer. The reason for this is that 80% of total mushrooms arise from the top layer of straw.

j. Put the last layer of 6 bundles and press the bed (Fig. 19).

k. Cover the bed with a transparent polythene sheet of 2m x 2m size. Tuck the edges of the sheet under the bed. Leave the bed as such for 10 days.

l. Remove the polythene cover after 10 days of sowing.

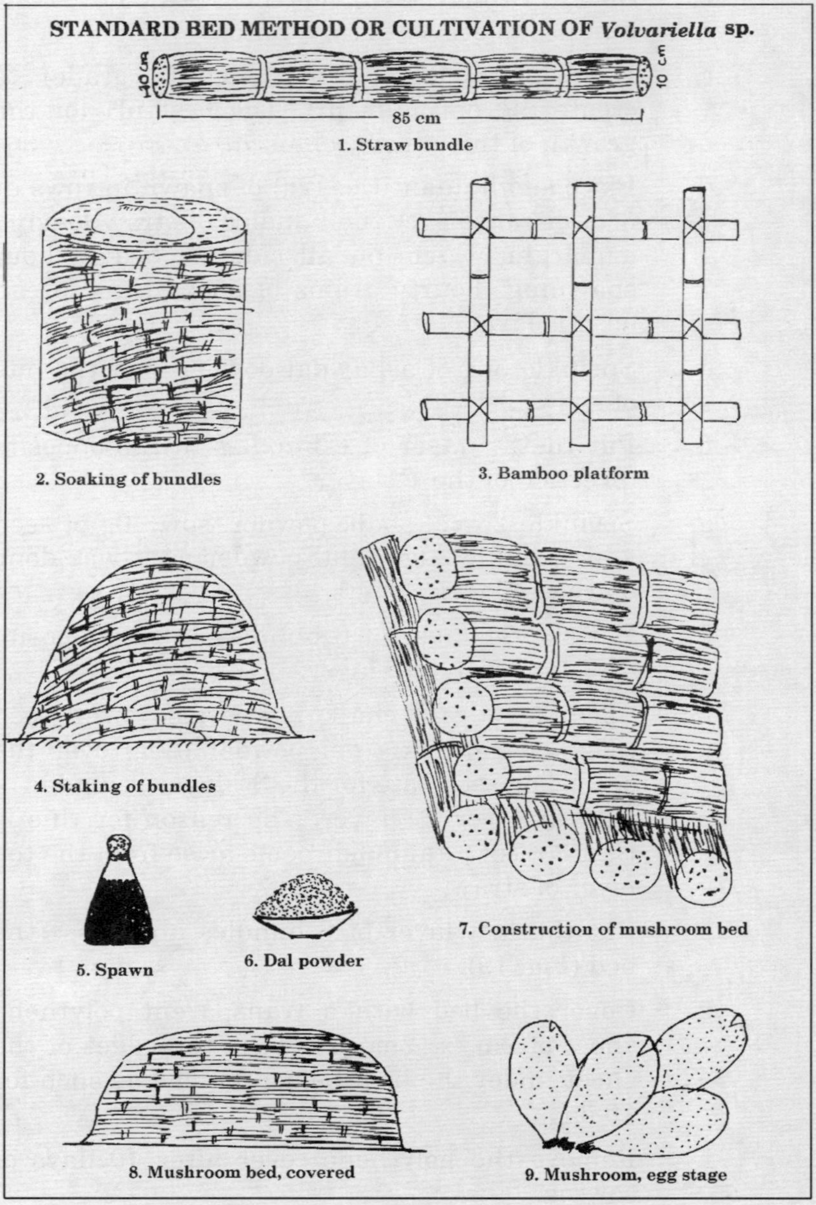

Fig. 17. Standard bed method for cultivation of *Volvariella* sp.

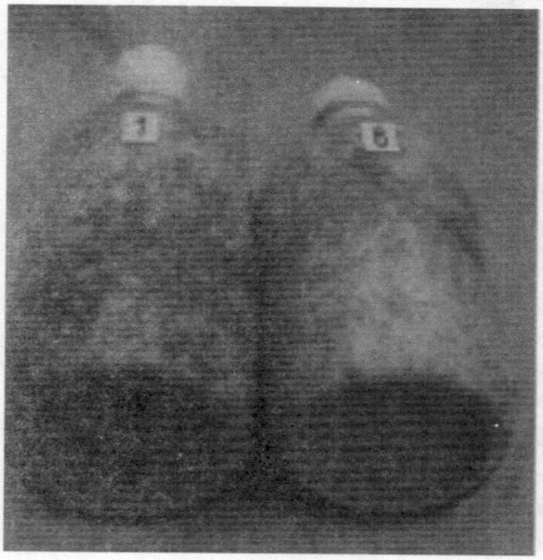

Fig. 18. Spawn of *Volvariella volvacea*

Fig. 19. Standard bed method

5.6.2. Hollow bed method

i. Make the straw into a rope of 7 cm diameter by twisting and tying knots at 15 cm interval. Four ropes each of 6 m in length, weighing 1.5 to 2 kg are required to construct one bed.

ii. Soak the straw ropes in water for 12 to 16h.

iii. Stack them into a heap to drain out excess water.

iv. Spray Ethion on straw just before spawning @ 2.5 ml/lt. to prevent mites.

v. A hollow cylinder of 30 cm diameter and 60 cm high was made by rolling a plain tin sheet.

vi. Place this cylinder vertically at the centre of the platform.

vii. Wind the straw rope around the cylinder. This becomes the 1st layer of straw.

viii. Sprinkle chalk powder, sow the spawn and add dal powder as described in the standard bed method.

ix. Place the 2nd, 3rd and 4th layers of straw twists planked by chalk powder, spawn and dal powder.

x. Remove the cylinder and cover the bed with polythene sheet for 10 days.

The advantage in hollow bed is that it is possible to make bed using a smaller quantity (8 kg) of straw.

5.6.3 Cage method (Fig. 20)

Materials required

i.	Paddy straw bundles	60/cage	vii.	Sprayer	1.no.
ii.	Spawn bottle	2/cage	viii.	Bavistin	1 pkt.
iii.	Wooden	1 no.	ix.	Formalin(500ml.)	1bottle
iv.	cage(1mx50cmx25cm)	1 no.	x.	Dao/Hachette	1 no.
v.	Drum (100 lt capacity)	4 m	xi.	Thermometer	1 no.
vi.	Polythene sheet	3 m	xii.	Arahar dal	200g
	Binding thread		xiii.	Turmeric powder	50 g

Methodology

i. Collect fresh dry and threshed paddy straw free from moulds. Make 25 cm long and 10 cm thick bundles of the paddy straw @ 60 bundles for each cage.

ii. Soak the bundles in water for 6-8 hrs and then immerse in boiling water for 20 – 30 minutes and allow to cool and drain out excess water.

iii. Disinfect the cage and polythene sheet with 4% Formalin solution.

iv. Arrange ten straw bundles uniformly in the cage as the bottom layer and put some spawn grains over it add little amount of arahar dal and turmeric powder. Repeat till six layers of bundles fill the cage.

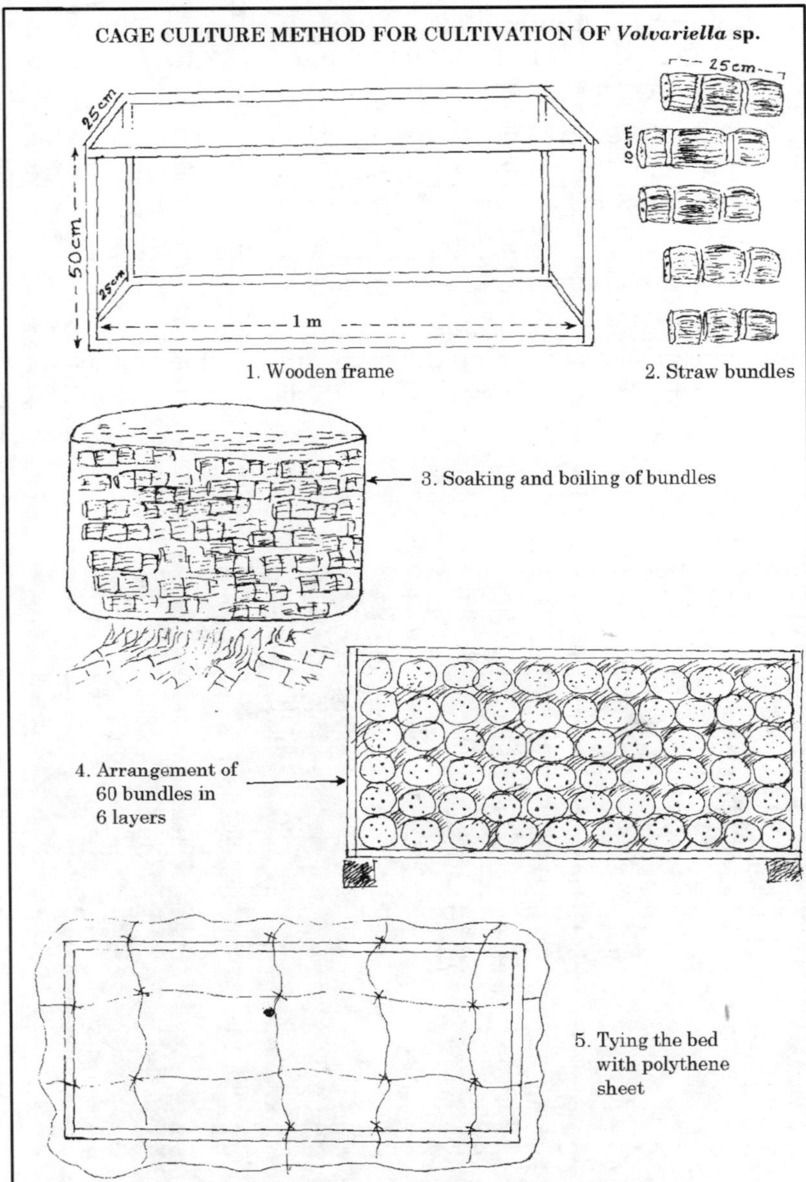

Fig. 20. Cage culture method for cultivation of *Volvariella* sp.

v. Spray solution of 0.2% Bavistin all over the bed. Cover with polythene sheet and tie securely with a binding thread.

vi. Keep the spawned cage in a room or a shed for mycelial run. A warm place with temperature around 30°C is required for better mycelial growth.

vii. Remove the polythene sheet after the mycelial run is completed (after 10 days).

viii. Maintain high humidity by watering with sprayer in the bed and inside the room till pin heads appear. Pin heads appear within 10-15 days after spawning (Fig. 21). Harvest mushrooms at the egg stage (Fig. 22).

ix. Continue spraying water for the next flush of mushroom to appear within a week or so.

Fig. 21. Paddy straw mushroom grown in cage method (appearance of pin heads)

Fig. 22. Egg stage and elongation stage of *Vovariella volvacea*

5.7. SUCCESSIVE GROWTH STAGES OF PADDY STRAW MUSHROOM

Sixth day: Silkworm's thread like mycelium is seen on all sides of the straw bed. This indicates that the spawning was successful.

Eighth day: Mycelial knots resembling sagu grains develop. These are called the primordia.

Ninth day: Tiny buttons of the size of pea grain are seen.

Tenth day: Mushrooms of button stage appear in compact cluster.

Eleventh day: Mushrooms attain egg stage. This is the best stage for picking.

Twelfth day: This growth stage is called elongation stage. The skin raptures and an unfolding umbrella-like mushroom push out. This is the next best stage for harvesting.

Thirteenth day: Mushrooms mature and looks like an open umbrella. These are highly perishable.

Twentieth day: A second crop of small quantity i.e. 250g/10kg straw comes up after a gap of 7 days.

The above time schedule for maturity of straw mushroom will vary according to the prevailing ambient temperature and humidity of the cultivation period.

5.8. PROBLEMS

i. Straw mites: Mites (*Tyrophagus dimidiatus*) are the most serious problem for paddy straw mushroom cultivation. They are of the size of sand grains. Innumerable mites are seen in mushrooms beds from 4 days of spawning. They devour the mycelium. Mites pose an irritating problem because they crawl on to body of person while harvesting mushroom. This leads to itching for several hours. Perfect control of mites is achieved by spraying Ethion 50EC @2.5 ml/lt. on straw one hour before spawning. The residue of Ethion in mushrooms harvested on 11th day of spawning is at safe level, i.e. 0.08mg/kg of mushroom.

ii. Weed mushroom: Ink cap fungus (*Coprinus* sp.) grows on the beds like weed that obstructs emergence of pin heads of paddy straw mushroom. This problem is very common on the beds of this mushroom. Management: Additional amount of $CaCO_3$ should be sprayed @ 10g/l on the beds after the first flush of mushrooms. Remove all the velvety young buds of *Coprinus* prior to their opening.

iii. Spawn availability: Spawn is not readily available. Hence, rice straw from partially spent beds, i.e. after harvesting the 1st flush of mushrooms is used as natural spawn in the following manner:

a. Spent straw is cut into pieces of 20 cm and used as spawn.

b. Four bundles, each of 1kg dry weight are used in constructing a new bed.

c. Yield is however lower (2.9 kg / 20 kg straw) as compared to that (3.1 kg / 20 kg straw) from grain spawn. Therefore, use natural spawn only when grain spawn is not available.

5.9. HARVESTING AND STORAGE

i. Harvest the fruit bodies at the egg stage. This mushroom harvested at full grown stage (umbrella stage) is less tasty.

ii. Do not store the fresh mushroom for more than a day. Otherwise fruit bodies will be watery because of its highly perishable nature. This mushroom should be consumed on the same day of picking.

5.10. TIPS FOR SUCCESS OF CULTIVATION

- Spawn: Use brick red coloured spawn.
- Cropping room: Raise the temperature and humidity inside the cropping room by putting an inner lining of plastic sheet.

CHAPTER VI

Cultivation of White Button Mushroom

6.1. SEASONAL CULTIVATION

White button mushroom (*Agaricus bisporus*) is a temperate mushroom (Fig. 23). It requires low temperature below 20^0C.The optimum temperature for fruiting is 18^0C. In Assam, its cultivation is done seasonally under natural conditions during winter months from November to January. The compost is prepared by long method. The seasonal units located in Haryana (Sonipat) and Punjab are the simple ways for the cultivation of this mushroom. This system can be adopted in Assam.

Haryana system

- The mushroom house is an ordinary mud hut.
- Compost is prepared by long method.
- Beds are prepared in shelf system made of bamboo frames.

- A polythene sheets is spread on bamboo shelves.
- Compost after spawning is filled in these shelves up to the depth of 4″ to 6″.
- One crop is consisting of 90 days duration.

Fig. 23. Button mushroom

6.2. CULTIVATION UNDER PREVAILING ENVIRONMENTAL CONDITIONS OF JORHAT

Only one date of sowing is possible

Compost : Preparation starts from 20th September

Spawning : 20th October to 1st November

Casing : 15th November

Crop : From 25th December to 20th January

6.2.1. Use of polythene bags in cultivation of button mushroom

Advantages

- There is no threat of contamination as new bags are used every time.

- Besides, chances of contamination by aerial microflora are minimum because bags are kept closed for first 15 days till the spawn run is completed.

  Size : 35″ x 24″ 150 gauge

  Capacity : 15 to 20 kg compost/ bag

  Depth : 30 to 37 cm

6.3. COMPOST

What is Compost?

A substratum in which mushroom mycelium grows and eventually develops mushroom fruit bodies. It is a selective medium for growing button mushroom. The process of compost preparation is called composting.

How is compost produced?

- It is a product of aerobic fermentation and decomposition process.
- The entire process is governed by a number of mesophilic & thermophilic organisms.
- During composting a change in chemical composition and physical state of the substrate takes place.

What is special about compost?

- Only *A. bisporus* can grow in it. Others will be excluded.
- Nutrients become readily available to the mushroom mycelia for synthesis of protein in the fruit bodies.

Quality of a good compost

A good quality compost should have the following

- Dark brown colour
- No greasiness but penetrable to air
- No smell of ammonia (maximum permissible limit is less than 10 ppm)

- Good water holding capacity but without water logging
- Appropriate pH ranging between 7.2 -7.5.
- Moisture 66-67%.
- Nitrogen content about 1.5-1.75%.

Important precaution

- Ammonia should not be present in the compost as it is inhibitory to mushroom mycelia.
- The moisture content of compost should not be more than 70%.

6.3.1. Materials required for compost preparation

i. Base material: Straw of wheat, paddy and barley. Of these wheat straw is the best.

Functions

- reservoir of cellulose, hemicelluloses, lignin and source of carbon.
- provides nitrogen (N of straw is <1% and hence, various inorganic and organic sources of N are supplemented to bring level of N to 1.5-1.75% in the compost).
- gives bulkiness and physical structure to the compost mixture.

ii. Carbohydrate rich material: Molasses, here N content is below 1.5%. The purpose of using this material is

- to correct C:N ratio (16:1).
- for establishing bacterial flora which can activate the fermentation process.

iii. Nitrogen rich materials: Ammonium sulphate, Urea, Calcium ammonium nitrate (CAN).

iv. Animal feeds: Wheat bran or rice bran is used for slow release of both N and carbohydrates. Their nitrogen content varies from 3-12%.

v. Materials for correcting deficiencies: These are Muriate of potash (MOP), Single super phosphate (SSP) and trace element.

vi. Gypsum: Its purpose is

- to precipitate suspended colloidal materials.
- to remove greasiness.
- for balancing pH.

6.3.2. Compost formula

A simple formula given by the National Research Center for Mushroom, Solan is:

Wheat straw or Paddy straw	1000 kg
Wheat bran	250 kg
Urea	18 kg
Gypsum	35 kg

A minimum of 300 kg straw is needed for making compost. When less than 300 kg straw is used, the required temperature inside the heap does not develop.

6.3.3. Methods of compost preparation

(i) Long method of composting: (Duration is 28 days)

Day-(-1) A. Wetting of straw: Compost can be made without chopping the straw. Clean the compost yard by washing the floor with 2% formalin. Spread the straw on the floor to 9″-12″ height, spray water, turn the straw and continue watering for 2-3 times. Leave as such for 24 hours. The moisture content of straw should be 75%.

B. Mix all the fertilizers except gypsum and rice bran, sprinkle water, make a heap, cover it with gunny bags and keep for 24 hours.

Day-0 Mix thoroughly all the fertilizers into wet straw. Stack the material into a pile (dimension 5ft x 5ft x 5ft, w x h x l) using a wooden mould (Fig. 24).

Moisture content of the straw should be 75-77%.

Press the compost mixture for compactness in the mould.

Mould is used for (i) obtaining a definite shape and height of the pile, (ii) aerobic fermentation to occur correctly.

Move the sideboards forward lengthwise, fill up with the mixture, press.

The purpose of mould is to obtain desired shape and height. More height of the pile leads to anaerobic fermentation while at low height sufficient heat will not be developed in the pile. Remove the planks after filling with straw.

Day-1-5 Leave the stack as such for 5 days. The temperature inside the pile goes up to 65-70^{0}C on the second day. This temperature is essential for getting good compost.

Day- 6 *First turning*: Break a portion of the pile e.g. 50 cm length; loosen it for exchange of air; water if there is a dry patch; erect the mould, fill in the mould and press from top.

Day- 10 *Second turning*: Break the heap from the rear; add $CaCO_3$ (if it is in the formula) during turning; water if required; make the pile, but pressing is reduced.

Day- 13 *Third turning*: Break the heap from the forward position; add gypsum (30 kg/ tonne) while turning; light pressing is required.

Day-16 *Fourth turning*: Turn; add Furadon 3G (400 g/ tonne) to check nematodes and re-stack. Cotton seed is added if it is in formula.

Day- 19 *Fifth turning*: Avoid lumps; pressing is not required.

Day- 22 *Sixth turning*: Ammonia smell should disappear or should be very low (less than 10 ppm); water the dry patches.

Day- 25 *Seventh turning*: spread the compost on a platform; break the lumps if any; mix BHC 5% dust (300g/tonne); stack in mould; water if requires.

Day- 28 *Filling stage*: Break the pile; check for ammonia smell; if it is present, give an additional turning after 3 days; Compost is ready for filling the trays or bags after spawning. The moisture content of compost should be 66-67%.

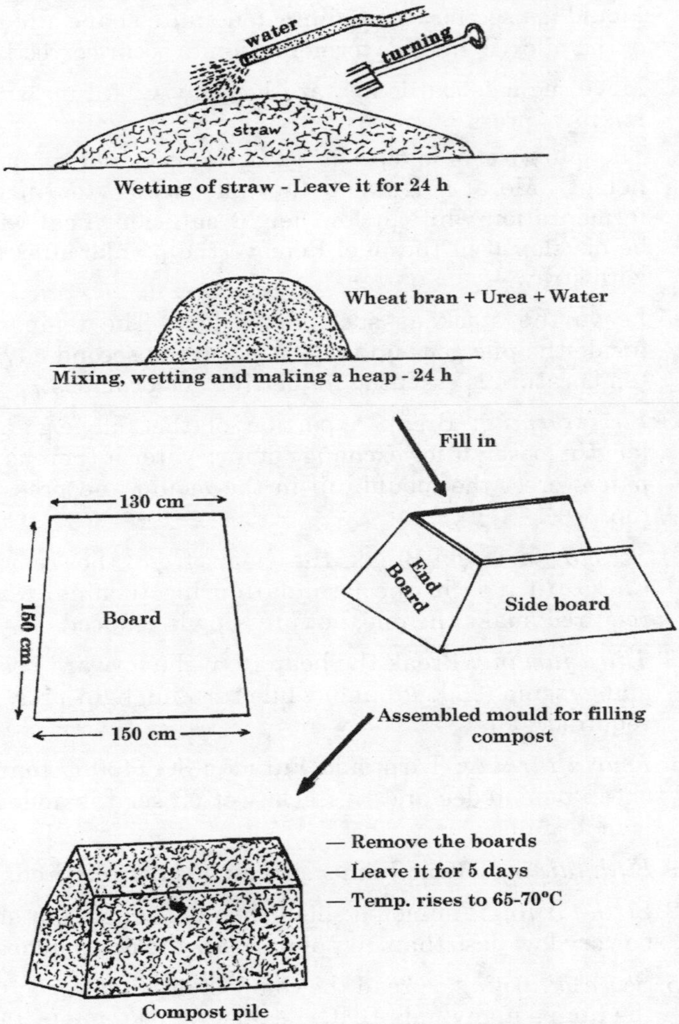

Fig. 24A. Flow chart for cultivation of Button Mushroom

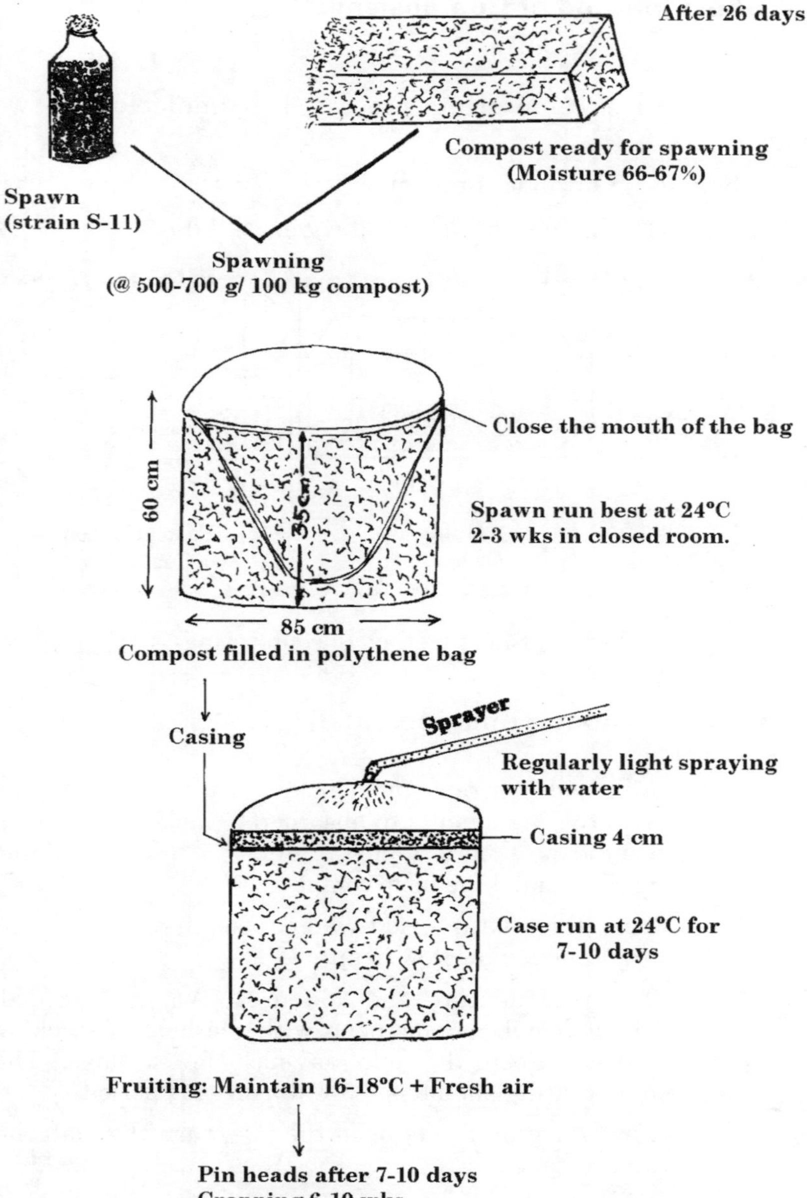
After 26 days
Compost ready for spawning
(Moisture 66-67%)
Spawn
(strain S-11)
Spawning
(@ 500-700 g/ 100 kg compost)
Close the mouth of the bag
60 cm
35 cm
Spawn run best at 24°C
2-3 wks in closed room.
85 cm
Compost filled in polythene bag
Casing
Sprayer
Regularly light spraying
with water
Casing 4 cm
Case run at 24°C for
7-10 days
Fruiting: Maintain 16-18°C + Fresh air
Pin heads after 7-10 days
Cropping 6-10 wks

Fig. 24B.

(ii) Short method of Composting: (Duration 18 days) Fig. 25

The entire composting process is completed in two phases:

Phase-I : Outdoor composting : 10-12 days

Phase-II : Inside an insulated room : 7 days

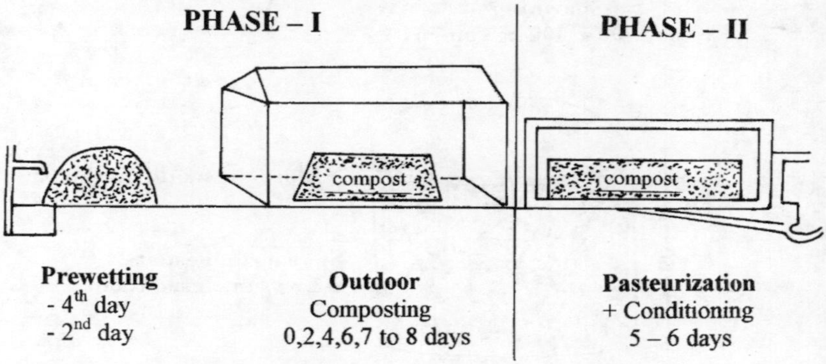

Fig. 25. Short method of composting

Phase I (Out doors; on open yard)

Day- (-4) Mix paddy straw and chicken manure thoroughly. Wet the ingredients to absorb 75% moisture. Make a low heap (3.3m x 2.5m x 114m, w x h x l). Trample hard to make the heap compact.

Day- (-2) Turn the mixture. Add water if required. Mix wheat bran and make a low heap.

Day- 0 Turn the mixture and add urea. Water if needed. Make a high aerobic heap with the help of stacking boards. Keep the pile as such for 2 days. The temperature should be 70^{0}C during this period.

Day- 2 *First turning*: Break open the pile; turn the material and restack.

Day- 4 *Second turning*: Break open the pile; add gypsum; turn the material and restack.

Day- 6 *Third turning*: Break open the pile; turn the material and restack.

Day- 8 Phase-I culminates. Manure is partially decomposed. Ready for Phase-II operation.

Phase-II (Indoor, in tunnel; Peak heating; Bulk pasteurization)

Day- 1 Fill the trays or tunnel with phase-I compost up to 7' feet height. Insert temperature probes in the compost. Hang one probe in the air. Shut doors and ventilators. Switch on the blower fan. Equalize the temperature of the air and compost at 45^{0}C (temperature should be brought to the required level by introducing fresh air and steam).

Day- 2-3 Introduce fresh air slightly by opening fresh air damper to 10% level. Maintain compost temperature between 45-52^{0}C for pre-pasteurization composting.

Day- 4 Raise temperature of compost mass to 59^{0}C by manipulating damper or by introduction of steam. Maintain this temperature for 6 h.

Day- 5-6 Maintain the temperature at 45-52^{0}C for post pasteurization conditioning for 2-3 days or till the period when compost is free from ammonia.

Day- 7 *Cooling*: Bring the temperature of compost to 25^{0}C by introduction of fresh air. The compost should be totally ammonia free.

Compost is ready on this day for spawning.

6.3.4. Difference between composting methods

Parameters	Long method	Short method
1. Duration required	28 days	16 to 20 days
2. Yield	Maximum yield 8-10kg/ 100kg compost	15-20 kg / 100 kg compost
3. Compost production	Less compost per unit wt. of ingredient	More compost per unit wt. of ingredient

4. Contamination	Liable to be infested by various insect pest and diseases	Free from diseases and pests.
5. Economy	Cheaper	Investment heavy
6. Type of composting	Completely outdoor method	Two phases: I: outdoor composting. II.: indoor: Pasteurization & conditioning

6.4. SPAWNING

Spawning is done @ 0.5-0.75% i.e. 500-750 g per 100 kg compost. Sowing of seed can be done by any one of the following three methods:

a. Surface spawning

i. Lift top 1-2″ layer of compost.

ii. Broadcast spawn.

iii. Ruffle with fingers into 1-2″ of compost.

iv. Cover with lifted compost and press uniformly.

b. Layer spawning

i. Spawn first layer at 3- 3.5″ depth.

ii. Put the next layer at 1-2″ from top.

iii. Cover thinly with compost.

c. Thorough spawning: Most efficient method.

i. Mix the spawn with the compost in a clean tub.

ii. Filling: Fill compost in polythene bags.

After spawning, compress the compost with hand or wooden plank in order to drive out air from the inner parts of the compost.

6.5. SPAWN RUNNING

Following spawning, a period of 14-18 days is required to complete mycelial run in the compost. Requirement of fresh air during this period is very low.

i. Cover the compost with news papers in case of trays or close the mouth of the compost filled poly bags by folding the ends of the bag.

ii. Spray water very lightly on the paper.

iii. Close the doors and windows.

iv. Maintain optimum temperature 25 ± 2°C and R.H. 90-95%.

v. Check compost for proper moisture. If surface is dry, give light water spraying.

Precaution: Temperature in the compost may go up after 24 h of spawning. Hence spray water on the floor and open the ventilators.

When the mycelium grows in the compost, the dark brown colour of the compost changes to light brown.

6.6. CASING

The process of applying a layer of soil over the compost after spawn run is called casing (Fig. 24 B). Casing layer stimulates fruit body formation. The casing soil should be poor in nutrients in order to create a stress condition for fruit body formation. It retains moisture and helps in gaseous exchange particularly CO_2. At 2% concentration of CO_2, mushroom mycelia grow well while fruiting occurs at lower level i.e. 0.08-0.15%. As a result of CO_2 stress condition, growth of the mycelia stops and fructification begins. Bacterium like *Pseudomonas putida* is present in casing material that plays role in the formation of pin heads. Casing layer provides a firm base for the mushroom fruit bodies. It regulates the flow of nutrients to the mushroom primordia from the compost.

Attributes of a good casing

i. It should have high water holding capacity (without water logging). Light soils are not good for casing. The casing soils should possess an apparent density between 0.75-0.81gl^{-1} and true density between 1.5-1.6 gl^{-1}.

ii. It should have good aeration capacity to remove unwanted gases during cropping.

iii. The pH should range between 7-7.5. Add lime or chalk powder @ 50g/kg soil to adjust the pH of casing material.

iv. It should contain enough iron.

v. Casing materials should be nutritionally poor.

vi. It should be free from insects, microorganisms, heavy metals and inorganic ions.

vii. Texture should not change due to watering.

6.6.1. Preparation of casing material

6.6.1.1. Casing formulae

i. Rotten cow dung + Soil (3:1)

ii. FYM + Loam soil (1:1)

iii. Two year old spent compost + Sand + lime (4:1:1)

iv. Soil + Moss (1: 1).

6.6.1.2. Pasteurization

(i) Steam pasteurization: It is done for 5-6 h at 60^{0}C or for 1½ h at 80^{0}C.

i. Fill in the trays with casing materials.

ii. Place the trays in an insulated chamber of 7′ x 7′ x 7′ size.

iii. Keep small gap of about 1- 1½″ between the trays.

iv. Fit perforated pipes at the bottom for uniform distribution of steam in the chamber.

v. Do steaming 24 h prior to spawning.

vi. The moisture content of casing soil should be 20%.

(ii) Chemical pasteurization: It is done by formalin treatment.

i. Apply chemical 2 weeks before spawning.

ii. 1000 kg of casing soil requires 1 lit of undiluted formalin.

iii. Wet the soil and add formalin.

iv. Cover the heap with polythene and leave as such for 3- 4 days.

v. Open the cover, allow the gases to go out.

vi. Rake the casing mixture after 2 days.

vii. Repeat raking every day up to 14 days

viii. Adjust pH at 7- 8.

6.6.2. Application of casing

- Apply the casing material of 3- 4cm thick when mycelial run is completed.
- Do not press the casing layer after application.
- Maintain 25^0C temperature for 8- 10 days after casing for faster growth of the mycelia in this layer.

6.6.3. Care after casing

Maintain the temperature of the spawn running room at 25^0C for 8 to 10 days for case run. The mushroom mycelium spreads into the casing layer. The casing soil should provide the moisture to the growing mushrooms. It also prevents drying of compost. Therefore, the casing soil must be kept in moist condition. Water should be sprayed in the form of a mist for 2 to 3 times in a day.

6.6.4. Cropping room

The cropping room should be built in a place where the direct sun rays do not fall. If sun- rays fall on the cropping room, it will be warmed up. There should be a provision for cross ventilation.

6.6.5. Cropping

The temperature inside the cropping room should be 14 to 18^0 C for inducing fruiting. This is a deciding factor for the success of a seasonal crop of button mushroom. If the temperature is above 18^0C during cropping, the yield will be reduced. There will be no mushrooms when the temperature is 25^0C or above. Good ventilation is needed for the mushroom development. The relative humidity should be 85 to 90%.

Under Assam situation, the preparation of compost should begin sometime between 15th to 20th of September. Then only, the required range of temperature for i. Vegetative growth, 25°C (spawn running) and ii. Fruiting 16°C will be met. If the starting of compost preparation is delayed up to middle of October, there will be no fruiting. Therefore, only one date of sowing is possible for button mushroom under natural conditions in Assam. The pinheads start appearing 12 to 15 days after casing. Mushrooms continue to come up for about 2 to 3 months. Under Assam situation, the button mushrooms come up during December and January. The strain S-11 of button mushroom performed better than other strains at Jorhat.

6.6.6. Harvesting

Mushrooms are harvested at button stage. The mushroom is held between the thumb and index finger and twisted slightly to detach it from casing soil. The base of mushroom is trimmed off. The holes formed in the casing soil after picking of mushrooms is filled with casing soil. Spraying of water is done only after harvesting of mushrooms.

6.6.7. Yield

An average yield of 10 to 14 kg fresh mushrooms per 100 kg compost can be obtained in a two months cropping period. Mushrooms can be stored in perforated polypropylene bags in a refrigerator for 7 days.

CHAPTER VII

Cultivation of Milky Mushroom

Calocybe indica is popularly known as milky mushroom. It is also called as "white summer mushroom" and "*Dudh Chatta*". This is a tropical mushroom. It was first reported from West Bengal as a wild edible species by R.P. Purkayastha and Aindrila Chandra in 1974 and its cultivation was started. Later on, a high yielding strain of *C. indica* was collected from a sugarcane field near Coimbatore in 1992. This strain was domesticated and released as a new variety of milky mushroom called APK-2 by Dr. A.S. Krisnamoorthy of Tamil Nadu Agricultural University (TNAU) in 1998. At present, Tamil Nadu is the leading state in commercial cultivation of milky mushroom. The cultivation technology of milky mushroom was perfected by Dr. A.S. Krishnamoorthy of TNAU, Coimbatore.

The work done at this University has proved that milky mushroom is successful in Assam. The cultivation technology, which is described here, is based on hands-on-experience.

7.1. SPECIAL CHARACTERS OF MILKY MUSHROOM

i. It can be grown at higher temperatures viz., 28 to 35°C.

ii. The keeping quality of this mushroom is good.

iii. This mushroom has a good texture which is retained after cooking.

iv. It requires casing similar to that in button mushroom.

v. It requires a special type of cropping room.

vi. It requires more light in comparison to oyster or button mushroom.

vii. It has the highest biological efficiency, up to 140%.

7.2. SEASON: May to September.

7.3. VARIETY: APK-2 (Fig. 26) from TNAU, Coimbatore.

7.4. STEPS OF CULTIVATION

Substrate: Wheat straw or paddy straw is the best substrates. Mushroom yield is higher when a blend of wheat straw + paddy straw (1:1w/w) is used.

Preparation of substrate: Straw is cut into small pieces of 3-5 cm size. It is soaked in cold water for 4hrs.

Hot water treatment: Water is boiled in a big aluminum sauce pan. The wet straw is immersed in hot water (80-90°C). The boiling of straw is continued for 1hr.

Sun drying: The boiled straw is spread on a clean cement floor. The straw is sprayed with a solution prepared with 10g $CaCO_3$ and 3- 5 ml garlic extract in 1 l water. Then the straw is allowed to dry for 30 min to 1 h depending on the prevailing weather. The moisture content of the straw should be brought down to 60 to 70%. Sometimes, the growth of the mycelium is inhibited after one week of spawning. This is the biggest problem in the cultivation of milky mushroom. About 20 to 50% of straw in the beds is colonized by the mycelium of *C. indica.* The remaining portion of straw is not colonized. A clear zone of inhibition is seen at the edge of the white patches of growth. There is no further colonization of the straw after one week of spawning. Rather ink caps (*Coprinus* sp.) start emerging from the uncolonised straw.

The major reason for the stoppage of mycelium of *C. indica* is believed to be the excess moisture content in the straw at the time of spawning. Therefore, correct moisture level in the straw is the crucial factor for the mycelial run of milky mushroom.

Fig. 26. Milky mushroom (Variety : APK-2)

Spawning: High density polypropylene bags of 30 cm x 60 cm are used. Twenty holes are made with a single punch machine. The bottom of the bag is tied with a thread to give a cylindrical shape to the bed. A spawn rate of 5% of the weight of wet straw is used. An increase in spawn rate i.e.

above 5% will give faster mycelial run and increase in yield. The bag is filled with alternate layers of straw and spawn. There will be 4 layers of spawn and 5 layers of straw. A ventilation pipe of PVC is placed at the centre of the bed at the time of filling. The beds are incubated at 28 to 30°C for spawn run.

Fig. 27. Mushroom house designed for milky mushroom.

Spawn running: The spawn run is completed in 15 to 20 days. Twenty days old spawn is better for spawn running and mushroom production. High light intensity of 1600 to 8000 lux favors fast growth of mycelium in beds. The growth of mycelium is slow in darkness. This is a contrasting character between milky mushroom and oyster mushroom.

Contamination of beds of *Rhizopus stolonifer* is a big problem when supplement like iron is used. Spray Apron SD (Metalaxyl) @ 0.5 g/l on sran before spawing, to control *R. stolonifer*.

Casing: Casing means covering the top surface of bags after spawn run with pasteurized casing material in a thickness of 4 cm.

Composition of casing material

i. One year old FYM + loam soil (1:1 w/w)

ii. Soil + sand (3:1 w/w)

iii. Soil + dried moss (1:1 w/w)

Besides the above, spent tea waste from tea stalls or ash from burnt rice husk can also be used for blending with casing soil. Moss is the best casing material because of its water holding capacity. Moss grows on walls in shady places. Chalk powder @ 2% (20 g/kg of casing soil) is mixed with the casing material in order to raise the pH to 8.4. The casing soil is pasteurized in an autoclave or in a pressure cooker at 15lbs/sq. inch for one hour. Moisture content of casing soil should be 50-60%.

Opening of beds: After completion of spawn run, the beds are cut into two halves with a hacksaw blade. Smaller bags of 20 cm in height, should not be cut into halves. Casing soil is applied to a height of 1-2 cm. above the newly exposed surface of the beds. The beds are kept in special type of cropping room.

Cropping room: The success of growing milky mushroom depends on the design of the cropping room. The Horticultural polyhouse can be converted into a cropping room for milky mushroom, with small modifications. In Assam, temperature during May to September remains below optimum for milky mushroom. The polyhouse is able to increase inside temperature above 30°C. The roof is made of UV stabilized low density polyethylene film. Netlon Agro shade of 75% is placed above the UV film in order to reduce the intensity of light. The roof is supported by brick wall on four sides (Fig. 27). Bamboo racks with 3-tiers are constructed inside the room. The size of the room is 4.5m in width and 6 to 10m in length. Milky mushroom requires light for spawn run as well as fruiting. Maximum yield of mushroom is obtained when there is 1600 lux light in the cropping room.

Pinning: The pinheads appear 6 to 8 days after casing. The mushrooms attain harvesting stage one week after pinning. A second and third crop of mushroom comes up at 10 days interval. A mist of water is sprayed every day on casing soil. The beds are retained up to 50 days after casing for mushroom production. Mushrooms should be harvested before the cap is fully opened.

7.5. SHELF LIFE

The shelf life of the mushroom is 5-7 days at room temperature. They can be stored in a refrigerator in perforated polypropylene bags for 20 days.

7.6. COOKING (RECIPES)

Milky mushroom has pungency. This is the only negative point. The pieces of mushrooms should be blanched by dipping in boiling water for 5-10 minutes and then rinsing with cold water. This process removes the pungency.

Curry

Ingredients

Milky mushroom – 250g	Onion – 2nos. big size
Tomato – 2 nos. big size	Turmeric powder – 1 tsp
Garlic paste – 1tsp	Chopped ginger – 1tsp
Cinnamon – 3 nos	Cardamom – 2 pieces
Oil – 3tbsp	Salt – to taste

Method

i. Chop the onion into pieces and make a paste.

ii. Cut the tomato into pieces and keep aside

iii. Heat the oil in a cauldron (*kadahi*), add the cinnamon and cardamom to it.

iv. Add the onion paste and fry it to golden brown colour at low flame. Pour the tomato pieces and fry till the oil comes out and the contents leave the sides of the kadahi.

v. Now add the mushroom pieces and cook at low heat for about 15 minutes.

Pickle

Ingredients

Milky mushroom – 200g
Onion paste – 20g
Garlic paste – 1tsp
Ginger paste – 1tsp
Chili powder – 1tsp
Cumin powder – 1tsp
Black pepper powder - ½tsp
Vinegar – 3tbsp
Mustard seed powder – 20g
Turmeric powder – 1tsp
Oil – 50 ml.
Salt – to taste

Method

i. Let the mushroom pieces dry for 1h after blanching.

ii. Heat oil in a cauldron (*kadahi*), fry the onion paste till its light brown colour. Add garlic and ginger paste and fry for sometime.

iii. Add cumin powder, chili powder, turmeric powder and the mushroom pieces and fry by stirring.

iv. Now add mustard powder, vinegar, salt and cook for 10 minutes.

v. Remove from heat and sprinkle black pepper powder.

7.7. PROBLEMS

1. Growth of mycelium is stopped after one week. Ink caps emerge from uncolonised straw. This is because

of high moisture content in straw. Straw should be dried on a cement floor in bright sun for 2 hrs. The moisture content should be 60-70%. The spawn rate should be increased to 6%.

2. Mother cultures die when stored in refrigerator. Cultures should be kept at room temperature and sub cultured at every 2 month interval.
3. Numerous ink caps of *Coprinus* sp. sprout from casing soil. These ink caps should be nipped everyday before they open. A light spray of formaldehyde @ 5ml/l is given over the surface of casing soil just after casing was done, in order to check *Coprinus* sp.

7.8. ECONOMICS OF MILKY MUSHROOM PRODUCTION

Milky mushroom grows well in the temperature range of 28 to 35^{0}C. It can be grown during the months of May to September. It needs a special mushroom house with plastic roof for increasing the inside temperature. One hundred days in a year are considered for production of milky mushroom. The same facilities can be utilized for growing oyster mushroom during the remaining period of a year. The economics of milky mushroom production has been with a capacity of 5 kg mushroom per day for 100 days in a year.

A. Fixed capital	**Amount (Rs.)**
1. Mushroom cropping room (poly house) of 3m x 6m with a capacity to hold 250 beds at a time	15,000.00
2. Spawn running room with tin roof, 3m x 6m with a capacity to hold 250 beds at a time	15,000.00
3. Chaff cutter, electrically operated	8,000.00

4.	Aluminum sauce pan for boiling of straw, 120 litres capacity	2,500.00
5.	Sprayers- two	400.00
6.	Fuel efficient chullah	500.00
	Total	41,400.00

B. Fixed cost

1. Interest @ 10% for 4 months of crop season	1,380.00
2. Depeciation on items 1 to 6 @ 10% for 4 months	1,380.00
Total	2,760.00

C. Working capital

1. Paddy straw, 0.5 ton	750.00
2. Spawn, 500 packets of 200 g each	5,000.00
3. Polythene bags	1,000.00
4. Labour, 100 man days @ 55/-	5,500.00
5. Fungicides	500.00
6. Miscellaneous	500.00
Total	13,250.00

D. Cost of mushroom production

1. Working capital plus fixed cost	16,010.00
2. Cost of production of 1 kg of mushroom (Rs.16,000/ 500 kg)	32.00

Steps in cultivation on Milky mushroom

Chop paddy straw into pieces of 3 to 5 cm

↓

Soak the straw in water for 6 to 8 hrs.

↓

Boil the straw in hot water (80°C) for 1 hr.

↓

Sun dry the straw for 1 to 2 hrs. or unitl 60-70% moisture level - a crucial step

↓

Spawn the substrate in layer cylindrical poly bags, 60 cm x 30 cm

↓

Incubate at 28°C for 15 to 18 days

↓

Cut the beds horizontally in to 2 halves

↓

Casing with soil + moss, 1:1, pH 8.5

↓

Keep the bags for fruiting at 28°-35°C in poly house with 75% shade net

↓

Pinheads formed in 8-10 days and ready for harvest in 5-8 days

↓

Yield 300-400 g/ 1/2 kg dry straw

E. Income

1. By sale of 5 kg mushroom/day @ 50/- for 100 days	25,000.00
2. Total cost of production	16,000.00
3. Net income out of 4 months	9,000.00

Supply of spawn

The spawn of milky mushroom is available from April to August in the Department of Plant Pathology, Assam Agricultural University, Jorhat-785 013, Ph: 0376-2341393. The price of spawn is Rs. 50/- per kg. The department also provides training course on cultivation technology of milky mushroom on request.

CHAPTER VIII

Some Edible and Inedible Fungi

EDIBLE FUNGI

I. GILL FUNGI

S. N.	Common Name	Botanical Name	Remark
1	Horse Mushroom	*Agaricus arvensis*	Edible
2	The Prince	*Agaricus angustus*	Excellent
3	Button Mushroom	*Agaricus bisporus*	Edible
4	Meadow or Meadow Mushroom	*Agaricus brunescens*	Edible
5	Field Mushroom	*Agaricus campestris*	Edible
6	Scaly Wood Mushrrom	*Agaricus sylvaticus*	Edible
7	Caeser's Mushroom	*Amanita caesarea*	Edible

8	—	*Amanita echinocephala, A. unbrinolutea*	Edible, poor flavour
9	Tawny Grisette Mushroom	*Amanita fulva (=A. vaginata* v. *fulva)*	Edible
10	—	*Amanita gemmata, A. inaurata, A. solitaria, A. spissa*	Edible with caution
11	Blusher	*Amanita rubescens*	Excellent cook ed, Toxic raw
12	Honey Fungus or Tuft	*Armillaria mellea*	Edible, Mild poisonous, needs parboiling
13	—	*Calocybe (= Tricholoma, Lyophyllum) ionides*	Good
14	St. George's Mushroom	*Calocybe gambosa (=Tricholoma georgii)*	Edible, excellent, highly prized
15	Chanterelle	*Cantharellus cibrius*	Edible, choice
16	Pig's Ears	*Cantharellus clavatus*	Edible, choice
17	Scaly Chanterelle	*Cantharellus floccosus*	Edible for some people
18	—	*Cantharellus cinnabarinus, C. lutescens, C. subalbidus*	Edible, choice
19	Orange Clitocybe	*Clitocybe aurantica*	Edible, but not recommended
20	Club Foot	*Clitocybe clavipes*	Edible
21	—	*Clitocybe (= Tricholoma, Lepista) glaucana, C. flaccida, C. langei*	Good
22	—	*Clitocybe geotropa*	Very good
23	Common Funnel Cap	*Clitocybe infundibuliformis*	Edible
24	Clouded Agaric	*Clitocybe nebularis*	Edible, mildly poisonous
25	Aniseed Toadstool, Anise Cap	*Clitocybe odora*	Edible

26 The Miller, Sweetbread mushroom	*Clitopilus prunulus*	Edible
27 Grassy or Butter Cap	*Collybia butyracea*	Edible
28 —	*Collybia dryophila, C. tenacella (= Strobilurus tenacellus), C. butyracea, C. fusipes*	Mediocre
29 —	*Collybia (=Maramius) confluens*	Edible
30 Common Ink Cap	*Coprinus stramentarius*	Edible, avoid alcohol
31 Shaggy Mane, Shaggy Ink Cap, Lawyer's Wig	*Coprinus comatus*	Edible
32 Glistering Ink Cap	*Coprinus micaceus*	Edible
33 Red banded Cortinarius	*Cortinarius armillatus, C. praestans*	Edible, good
34 Trumpet of Death or Horn of Plenty	*Craterellus cornucopioides*	Edible
35 —	*Cystoderma amianthinum*	Edible, mediocre
36 —	*Cystoderma carcharias*	Edible, poor quality
37 Velvet stem, Winter fungus, Enok-itake or Enoki Mushroom	*Flammulina (=Collybia) velutipes*	Edible
38 — 39	*Hygrocybe brevispora (= Hygrophorus brevisporus), H. conica (Hygrophorus conicus), H. psittacina (Hygrophorus psittacinus)*	Mediocre
40 —	*Hygrophorus limacinus, H. caprinus*	Very good
41 False Chaterelle	*Hygrophoropsis aurantiaca*	Edible

42 Conifer Sulphur Tuft	*Hypholoma capnoides*	Edible
43 Deciever	*Laccaria laccata*	Edible
44 Weeping Widow	*Lacrymaria velutina*	Edible
45 Delicious Lactarius	*Lactarius deliciosus*	Edible, choice
46 Saffron Milk Cap	*Lactarius determinus*	Edible
47 Coconut Milk Cap	*Lactarius glycimosus*	Edible
48 Oily Milk Cap	*Lactarius quietus*	Edible
49 —	*Lactarius sanguifluus*	Excellent
50 Shiitake or Black Forest Mushroom	*Lentinus edodes*	Edible
51 Scaly Lentinus	*Lentinus lepideus*	Edible
52 Parasol Mushroom	*Lepiota (=Macrolepiota) procera*	Edible
53 —	*Macrolepiota (=Lepiota) puellaris*	Edible, good
54 Shaggy or Woodland Parasol	*Lepiota (=Macrolepiota) rhacodes*	Edible
55 Wood Blewits	*Lepista nuda (=Clitocybe nuda, Tricholoma nudum)*	Excellent, slightly poisonous raw, needs parboiling.
56 Common field Blewit, Blue Leg	*Lepista saeva*	Edible
57 —	*Leucoagaricus pudicus*	Edible with caution
58 —	*Leucopaxillus giganteus (Clitocybe gigantean)*	Very good

59 —	*Limacella guttata* (*=L. lenticularis, Lepiota lenticularis*)	Edible
60 Garlic Mushroom	*Marasmeus alliaceous, M. scorodonius*	Used for flavouring or seasoning
61 Fairy Ring Fungus	*Marasmeus oreades*	Edible
62 Orange Bonnet	*Mycena acicula*	Edible
63 Lilac Mycena	*Mycena pura*	Edible
64 Brick Cap/ Stump mushroom	*Naematoloma sublateritium*	Edible
65 -	*Oudemansiella (=Collybia) radicata*	Edible, mediocre
66 Shaggy or Scaly Pholiota	*Pholiota squarrosa*	Edible, but avoid old specimens
67 Nameko	*Pholiota nameko*	Good
68 Oyster Mushroom	*Pleurotus citrinopileatus, P. cornucopie, P. dryinus, P. eryngii, P. flabellatus, P. florida, P. ostreatus, P. sapidus*	Edible, excellent
69 Grey Oyster	*Pleurotus sajor-caju*	Edible
70 —	*Psilocybe squamosa (= Stropharia squamosa, Naematoloma squamosam)*	Edibility uncertain
71 Porcelain Fungus	*Oudemansiella (= Armillaria, Mucidula) mucida*	Edible
72 Rooting Shank	*Oudemansiella radicata*	Edible
73 —	*Stropharia aeruginosa*	Edible
74 —	*Stropharia coronilla*	Edible, mildly toxic
75 Brown Cap, Giant Mushroom	*Stropharia rugosa-annulata (=Stropharia ferrii)*	Edible, delicious
76 Termite nest fungus	*Termitomyces hemii*	Edible

77	—	*Termitomyces mammiformis, T. microcarpus*	Edible
78	—	*Tricholoma ageratum (= Lyophylum decastes, Clitocybe multiceps), T. flavovirens (= T. equestre), T. portentosum, T. terreum*	Very good
79	—	*Tricholoma cnista (= Melanoleuca cnista, M. evenosa)*	Edible
80	—	*Tricholoma caligatum (= Armillaria caligata)*	Excellent
81	—	*Tricholoma imbricatum, T. lascivum, T. ustaloides*	Mediocre
82	—	*Tricholoma sejunctum*	Edible, not tasty and pleasant
83	Paddy Straw Mushroom	*Volvariella diplasia, V. volvacea, V. bombycina, V. esculenta*	Edible

II. BOLETES AND ALLIES

1	Bay-capped Bolete	*Boletus badius*	Edible
2	Red-cracked Bolete	*Boletus chrysenteron*	Edible
3	Cepe / Penny Bum Mushroom/ Steinpilz	*Boletus edulis*	Edible, excellent
	—	*Boletus aurantiacus, B. aereus, B. chromapes, B. mirabilis, B. variipes*	Excellent
4	Downy Bolete	*Boletus subtomentosus*	Edible
5	Yellow cracking Bolete	*Leccinum crocipodium*	Edible
6	Brown Birch Bolete	*Leccinum scabrum*	Edible
7	Old man of the Woods, Pine cone Bolete	*Strobilomyces floccopus*	Edible

8	Larch Bolete	*Suillus grevillei*	Edible
9	Slippery Jack	*Suillus luteus*	Edible

III. BRACKETS AND ALLIES

1	Club Fungi	*Clavaria argillacea*	Edible
2	Black Trumpet or Horn of Plenty	*Craterellus cornucopiodes*	Edible
3	Beef Steak, OK tongue	*Fistulina hepatica*	Edible
4	Artists Fungus	*Ganoderma curopaum*	Too hard to eat
5	Wood Hedge Hog	*Hydnum repandum*	Edible
6	Sulphur Polypore or Sulphur Shelf	*Lactiporus(=Lateiporus) sulphureus*	Edible
7	—	*Polyporus betulinus (=Piptoporus betulinus), P. griseus (=P. leucomelus, Boletopsis subsquamosa), P. ovinus (= Albatrellus ovinus), P. umbellatus*	Edible
8	—	*Polyporus pes-caprae (=Albatrellus pes-caprae)*	Edible, excellent
9	Dryad's Saddle or Dry scaly Polypore	*Polyporus squamosus*	Edible
10	Canadian Tuckahoe	*Polyporus tuberaster*	Edible
11	Cauliflower Fungus	*Sparassis crispa*	Edible

IV. JELLY FUNGI

1	Woody Ear, Black Ear, Tree Ear	*Auriculria auricula*	Edible

2	Jew's Ear or Judas' Ear	*Hirneola auricula-judae*	Edible
3	Jelly Tongue	*Pseudohydnum gelatioasum*	Edible
4	White Jelly Fungus or Silver Ear	*Tremella fuciformis*	Edible

V. PUFF BALLS AND ALLIES

1	—	*Calbovista subsculpta*	Edible, popular
2	Giant Puffball	*Calvatia gigantea*	Edible, popular
3	Common Puffball	*Lycoperdon perlatum (= L. gemmatum)*	Edible
4	Stump or Pear-shaped Puffball	*Lycoperdon pyriforme*	Edible
5	Common Stink Horn	*Phallus impudicus*	Edible

VI. CUP FUNGI

1	Common white Helvella	*Helvella crispa*	Edible
2	Black or Grey Helvella	*Helvella lacunosa*	Edible
3	Morel	*Morchella conica*	Edible
4	Common Morel	*Morchella esculenta*	Edible
5	—	*Morchella elata, M. umbrina, M. vulgaris*	Excellent
6	Truffle	*Tuber aestivum*	Edible
7	Red Truffle	*Tuber rufum*	Edible
8	Thimble Fungus	*Verpa conica*	Edible

VII. MORELS AND FALSE MORELS

1	Black Morel	*Morchella angusticeps*	Excellent, choice
2	Thick-Footed Morel	*Morchella crassipes*	Excellent, choice
3	Morel or Sponge Mushroom	*Morchella esculenta*	Excellent, popular
4	Half-Free Morel	*Morchella hybrida*	Edible, good flavour
5	Early Morel	*Verpa bohemica*	Not advisable to take large amount
6	Snow Mushroom	*Helvella gigas*	Edible, choice

NON-EDIBLE FUNGI

I. GILL FUNGI

1	Yellow Stains	*Agaricus xanthodermus*	Poisonous
2	—	*Amanita citrina*	Slightly poisonous when raw
3	Fly Agaric	*Amanita muscaria*	Non- edible
4	American Fly Agaric	*Amanita muscaria formosa*	Poisonous
5	Panhter Cap	*Amanita pantherina*	Dangerous and fatal
6	Death Cap (= Green Death Cap)	*Amanita phalloides*	Deadly poisonous
7	Fool's Mush-room, Spring destroying Angel	*Amanita verna*	Deadly poisonous

8	Destroying Angel	*Amanita virosa*	Non-edible
9	Jack-O-Lantern Fungus	*Clitocybe illudens*	poisonous
10	—	*Clitocybe phyllophila*	Poisonous
11	Spindle Stem	*Collybia fusipes*	Non-edible
12	Red spot Cap	*Collybia maculata*	Non-edible
13	—	*Collybia peronata (= Marasmius peronatus)*	Non-edible, but used as flavouring agent
14	Wood Wodly Foot	*Collybia personata*	Non-edible
15	—	*Coprinus (= Pseudocoprinus) disseminatus, C. niveus, C. plicatilis C. radians, C. silvaticus*	Of no interest because of small size
16	—	*Coprinus picaceus*	Slightly poisonous
17	—	*Cystolepiota adulterine*	Non-edible
18	Livid Entoloma	*Entoloma sinuatum*	Poisonous
19	Poison Pie	*Hebeloma crustuliniforme*	Poisonous
20	Sulphur Tuft	*Hypholoma fasiculare*	Non-edible
21	Brick Cap, Red Cap	*Hypholoma sublateritium*	Non-edible
22	Common White Inocybe	*Inocybe geophylla*	Poisonous
23	Slimy Milk Cap	*Lacterius blennius*	Non-edible
24	Woody Milk Cap	*Lacterius torminosus*	Non-edible
25	Stinking Parasol	*Lepiota cristata*	Possibly poisonous
26	—	*Lepiota friesii (=L. acutesquamosa)*	Non-edible

27	—	*Marasmius foetidus (=Microphale foetidum)*	Non-edible, terrible flavour
28	—	*Marasmius(=Marasmeallus)ramealis, M.(=Microphale)perforans, M. rotula*	Of no value for size
29	Jack-o' Lantern	*Omphalotus olearius*	Poisonous
30	Charcoal Phaliota	*Pholiota highlandensis*	Non-edible
31	Liberty Cap	*Psilocybe semilenceata*	Non-edible
32	Dung Roundhead	*Stropharia semiglobata*	Of no interest because of size
33	Soap Scented toadstool	*Tricholoma saponaceum*	Non-edible
34	Sulphur Toadstool	*Tricholoma sulphureum*	Non-edible
35	—	*Tricholoma pardinum, T. virgatum*	Poisonous

II. BOLETES AND ALLIES

1	Miss Alice Eastwood's Boletus	*Boletus eastwoodiae*	Poisonous
2	Peppery Bolete	*Boletus piperatus*	Non-edible
3	Devil's Bolete, Satan's Bolete	*Boletus satanus*	Non-edible
4	Brown Roll Rim	*Panilus involutus*	Non-edible

III. BRACKETS AND ALLIES

1	Earpick Fungus	*Auriscalpium vulgare*	Non-edible
2	Silver Leaf Disease Fungus	*Chondrosterum purpureum*	Non-edible

3	Golden Spindles	*Clavullinopsis fusiformis*	Non-edible
4			
5	Tinder or Hoof Fungus	*Fomes fomentarius*	Non-edible
6	Artist's Fungus	*Ganoderma applanatum, G. europeum*	Too hard to eat
7	Lacquered Bracket, Reishi Mushroom	*Ganoderma lucidum*	Non-edible
8	Shaggy Polypore	*Inonotus hispidus*	Non-edible
9	Birch Polypore	*Piptoporus betulinus*	Non-edible
10	Polypore Mushroom	*Polyporus brumelis*	Of no interest
11	—	*Polyporus mori (Favolus alveolaris, F. Canadensis, F. europeus)*	Of no interest
12	—	*Polyporus (=Polyporellus) arcularius, P. (=Coltricia) perennis, P. schweinitzii*	Non-edible
13	—	*Polyporus cinnabarinus (Pycnoporous cinnabarinus)*	Very tough, inedible
14	—	*Polyporus tomentosus (= Onnia tomentosa)*	Too leathery, inedible
15	Coral Fungi	*Ramaria botrytis*	Non-edible
16	Earth Fan	*Thelephora terrestris*	Non-edible

IV. JELLY FUNGI

1	-	*Ascocoryne sarcoides (= Coryne sarcoides)*	Of no interest
2	-	*Auricularia mesenterica*	Non-edible
3	Yellow Staghorn Fungus	*Calocera viscosa*	Non-edible

4	Jelly Tongue	*Pseudohydnum gelatinosum*	Edible but not recommended
5	Golden Jelly or Yellow Brain Fungus	*Tremella mesenterica*	Non-edible

V. PUFFBALLS AND ALLIES

1	Cage Fungus	*Clathrus ruber*	Non-edible
2	-	*Cyathus olla*	Of no interest
3	Bird nest Fungus, Splash cup	*Cyathus striatus, Crucibulum vulgare*	Non-edible
4	Earth Star	*Geastrum striatum, G. triplex*	Non-edible
5	Spiny Puffball or Hedgehog	*Lycoperdon echinatum*	Non-edible
6	Common Earthball	*Scleroderma citrinum*	Non-edible

VI. CUP FUNGI

1	Orange Peel Fungus	*Aleuria aurantica*	Non-edible
2	Rubber Button	*Bulgaria inquinans*	Non-edible
3	Green Wood Cup	*Chlorociboria aeruginascens*	Non-edible
4	Black or Grey Helvella	*Helvella lacunosa*	Edible but not recommended
5	False Morel	*Morchella esculenta*	Non-edible
6	Hare's or Rabbit Ear	*Otidia leporina*	Non-edible
7	Scarlet Cup or Elf Cup	*Sarcoscypha coccinea*	Non-edible
8	Eye Lash Fungus	*Scutellinia scutellata*	Non-edible

VII. FLASK FUNGUS

1	King Alfred's Cakes, Crampball	*Daldinia concentrica*	Non-edible
2	Coral Spot	*Nectria cinnabarina*	
3	Candle Snuff Fungus	*Xylosphaera hypoxylon*	Non-edible
4	Dead Man's Finger	*Xylosphaera polymorpha*	Non-edible

Some poisonous and edible species of mushroom commonly available in Assam:

Sl. No.	Scientific name	Habitat
A.	**Poisonous**	
1.	*Amanita phalloides*	Grasslands and decomposed debris
2.	*Amanita muscaria*	-do-
B.	**Edible**	
1.	*Agaricus campestris, A. squamaulferous*	Ploughed soil
2.	*Armillaria mellea*	Tree stumps
3.	*Calocybe* sp. (*Tricholoma gambosum, T. lascivum*)	Grasslands and shady moist soil
4.	*Calvatia utriformis*	Grass land
5.	*Cantherellus brevipes, C. cibarius*	Tree stump
6	*Clitocybe flaccida*	Felled tree
7	*Coprinus stramentarius*	Soil around tree
8	*Hirneola auricula- judae*	Tree stump
9	*Hygrophorus citrinus, H. conicus*	Grass land and dry sandy soil
10	*Lacterius* sp.	Tree stump
11	*Lepiota leucothites, L. procera* (Fig. 28)	Decomposed cow dung, debris, soil.
12	*Lycoperdon pocelum, L. pyriforme, L. pusillum*	Ploughed soil

13	*Marasmius oreades*	Grass land and weed free soil
14	*Pleurotus citrinopileatus, P. squarrosulus*	At the base of trees, tree stumps
15	*Schizophyllum commune*	Rotten wood
16	*Termitomyces clypeatus, T. heimii* (Fig. 30), *T. eurrhizus, T. robustus*	Termite nest, sandy soil
17	*T. microcarpus, T. mammiformis*	Sandy soil after rain

IDENTIFICATION OF SOME COMMON WILD EDIBLE MUSHROOMS OF ASSAM

1. *Macrolepiota procera* syn: *Lepiota procera*

This mushroom attracts the attention of passers-by because of its tall (up to 25 cm) stipe and a large sunshade like cap (up to 17 cm diameter). Therefore, it is called as 'Parasol mushroom' or the 'tall Lepiota' (Fig. 28). This mushroom is found in several places in India viz. Assam, U.P., Andhra Pradesh, Himachal Pradesh and Kashmir.

Fig. 28. *Lepiota procera*

Season: June is the best month for this mushroom. The maximum and minimum temperature during fruiting are 30° and 24°C respectively. Rain always precedes the emergence of the mushroom.

Edibility: It tastes good. People pluck it when the cap is fully open. The stalk is fibrous. The quantity and quantity of mushroom after cooking is very much reduced. Mushrooms plucked during rainy weather are tastier than those picked during bright sunny days as desiccation occurs due to dry condition. Its protein content is 32% on dry weight basis.

Identifying features

i. Stalk is tall, slender resembling a jute stick after retting. It is coated with pink granules.

ii. A large cap with rusty brown coloured umbo (projection).

iii. Spore print is white.

iv. A movable and prominent ring.

Technical description: Pileus 14–16 cm in diameter, ovate when young, becoming companulate, convex and finally nearly plane with a low obtuse umbo covered at first with a reddish brown cuticle which ruptures to form flattish scales as the cap expands; lamellae free, remote, white, becoming faintly pinkish; stipe 20-25 cm long, 8-12 mm thick at apex, base bulbous, separating readily from the pileus, hallow, furfuraceous, annulus superior and movable (Fig. 29 a,b,c,d,e) spores ellipsoid, 8-12 x 5-8 μm (Fig. 29f), white in mass with a thick wall; basidia 4-spored.

Parasol mushroom was cultured on PDA. Desiccated pellets of cattle dung is the preferred substrate for this mushroom. Its spawn was prepared in rice and wheat grains. Spawning was done in pasteurized paddy straw, compost made from cattle dung as well as synthetic compost.

There was no fruiting in any of the substrates even though mycelial run was luxuriant in paddy straw after 15 days of spawning. Jandaik and Thinga (1980) from Himachal Pradesh, artificially cultivated this mushroom using synthetic compost as prepared for button mushroom. This mushroom is not easy to domesticate for commercial production.

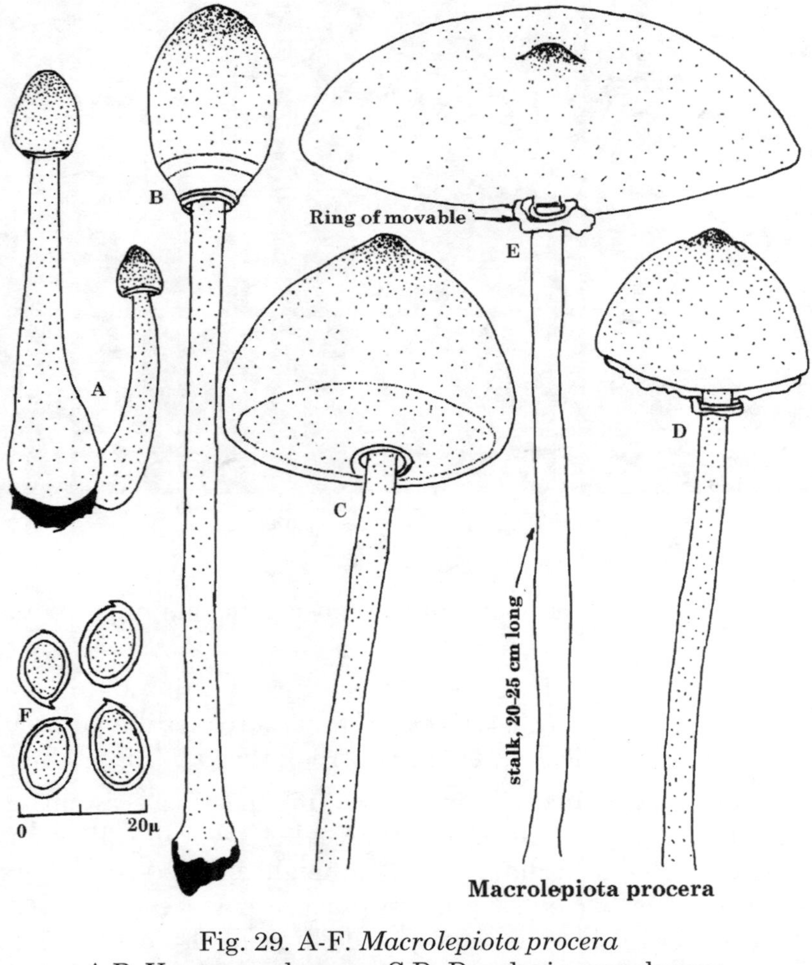

Fig. 29. A-F. *Macrolepiota procera*
A,B. Young mushroom, C,D. Developing mushroom,
E. Fully opened mushroom

2. *Termitomyces clypeatus*

This is the most dominant species among all the wild edible mushrooms of Assam (Fig. 30). This mushroom comes up from termite mounds. It is found in North Eastern states and South India.

Fig. 30. *Termitomyces clypeatus*

Season: June to August.

Edibility: This is one of the most delicious mushrooms found in nature.

Identifying features: The most distinguishing character of this species is the prominent, cone-like pointed, dark coloured projection at the centre of the cap (Fig. 31E)

Technical description: The pileus is 4-8.5 cm in diameter. The largest one among the several collections measured 13 cm in diameter. The height of the stalk above the ground is 3-7 cm. Chilocystidia are clavate (Fig. 31G). The spore print is pink. Spores are ellipsoid, 4.5-6 x 3-4.5μm (Fig. 31F).

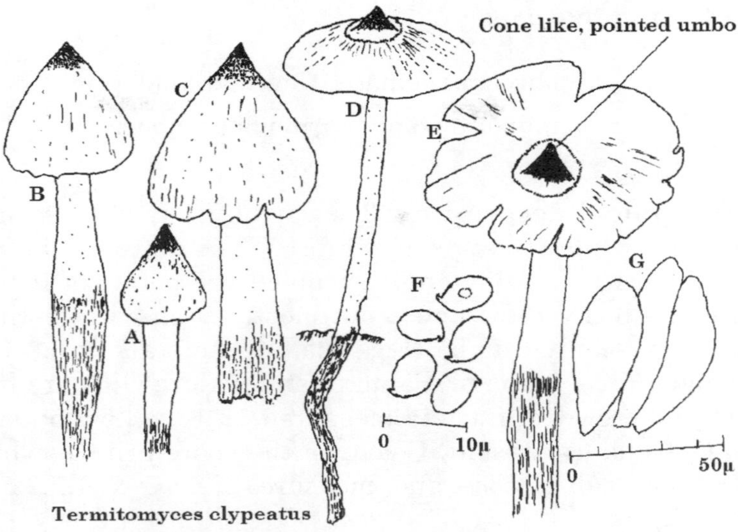

Fig. 31. A-F. *Termitomyces clypeatus*
A,B,C. Young mushrooms, D,E. Mature mushroom
F. Spores, G. Chilocystidia

3. *Termitomyces microcarpus*

This is one of the most prevalent wild edible mushrooms found in Assam which is locally called as 'soru naphu' i.e. a tiny Termitomyces. This mushroom grows profusely in bamboo grove from beneath fallen bamboo leaves. It is always associated with colonies of ants and anthills. This mushroom always comes up in several hundreds in a relatively small place. They resemble wild flowers. They look like a matchstick, when first seen above ground. It is found in Assam, West Bengal, Tamil Nadu and Karnataka.

Season: July is the best month for this mushroom. The minimum and maximum temperatures during this period prevail are 22.5° and 29.8°C, respectively.

Edibility: It is one of the bests in taste. After cooking, the mushrooms become pulp like and give a minched-meat like taste.

Identifying features

i. Size of mushroom is small, 2 cm in diameter.

ii. It comes up in crowd of enormous number.

iii. Comes up mostly from colonies of ants.

Technical description: The cap of young mushroom is light grey while its stalk is white. When mature, the cap splits from its margins and resembled the petals of flower. Cap is 2 – 6 cm in diameter, but mostly 2 cm with distinct, light grey coloured umbo (Fig. 32a,b). The stalk is 6.5 to 8 cm long and 2 – 3 mm in diameter and hollow. Gills are free, white and crowded. Basidia measured 12-18 x 4.5-7.5 μm in size. Cystidia are absent. Colour of the spore print is white. Spores are sub-globose and measure 12-18.5 x 4.5-7.5μm (Fig. 32E).

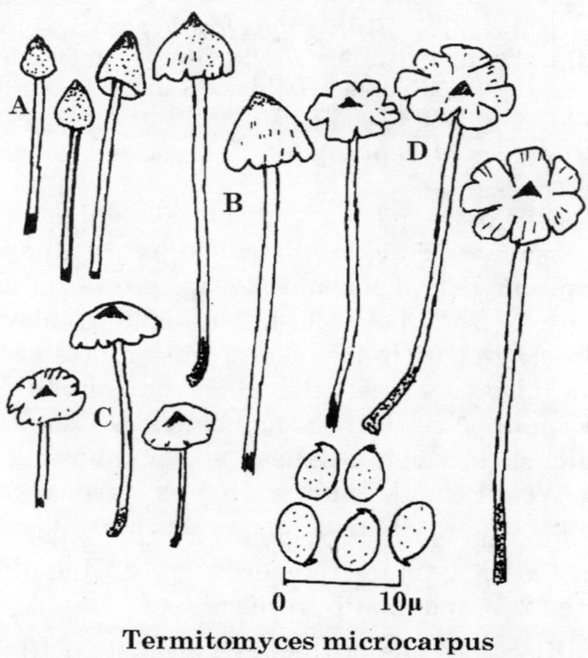

Termitomyces microcarpus

Fig. 32. A-E. *Termitomyces microcarpus*
A,B. Immature mushroom, C,D. Mature mushroom, E. Spores

4. *Termitomyces heimii*

This mushroom comes up exclusively from termite nests (Fig. 33). The local name is 'Bor Naphu' i.e. big Termitomyces. It is found in Assam, West Bengal, Tamil Nadu, Karnataka and Rajasthan.

Season: July to August.

Edibility: Very delicious.

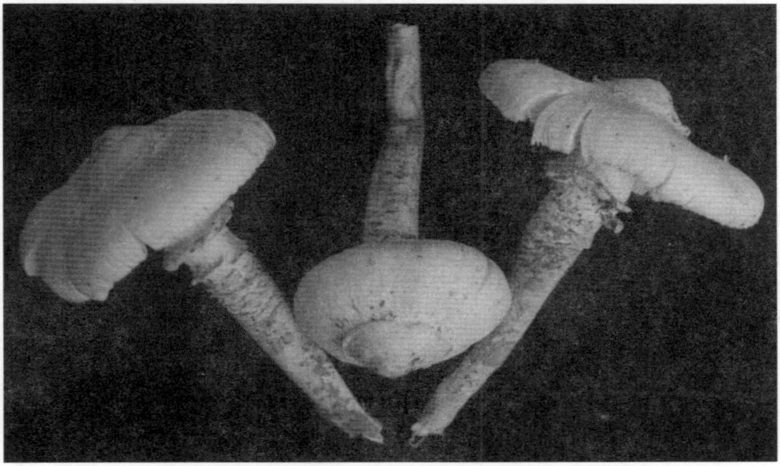

Fig. 33. *Termitomyces heimii*

Identifying features

i. A patch of veil is always present at the top of the pileus (Fig. 34a).

ii. The skin of the cap peels off.

iii. The stalk is long (up to 25 cm) and swollen near the attachment with the cap.

Technical description: Pileus is 8-10 cm in diameter, white, sub-unbonate with a patch of veil at the apex. The cap splits at a few places from margin. Stipe is 25 cm long on an average, bulges near the pileus and possess a thick ring. Pleurocystidia are clavate (Fig. 34b). Spore print is deep pink. Spores are ellipsoid and measure 4.5-6 x 7.5-9μm (Fig. 34c).

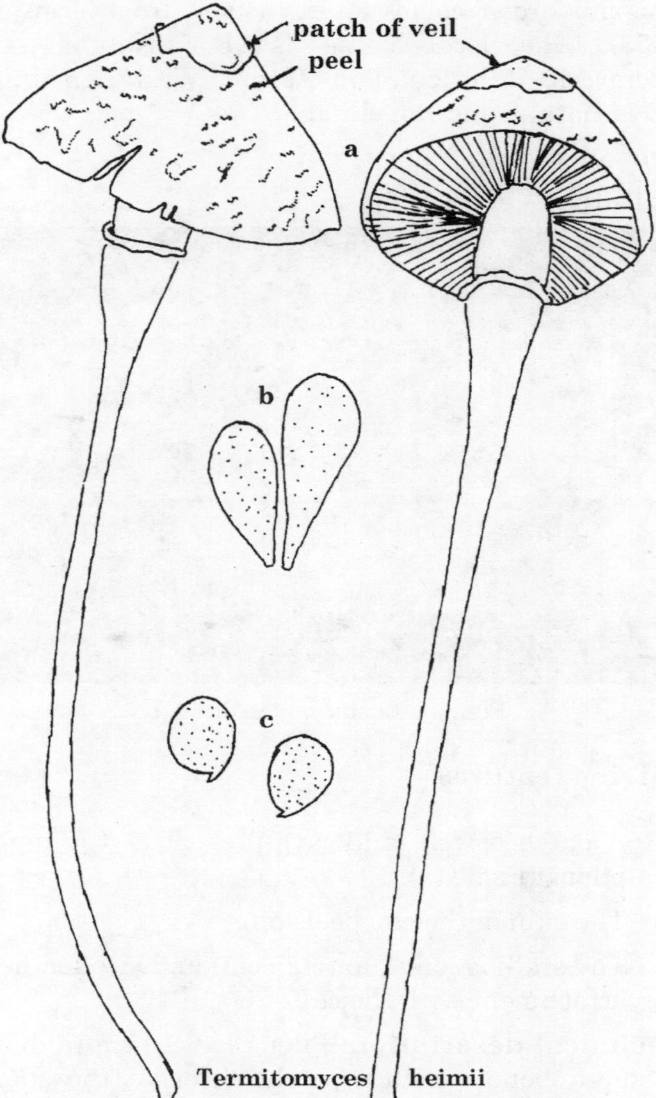

Fig. 34. a-c. *Termitomyces heimii*
a. Mushrooms, b. Pleurocystidia, c. Spores.

CHAPTER IX

Production of Vermicompost from Spent Mushroom Beds

Spent mushroom beds of oyster and paddy straw mushroom are the best substrate for making vermicompost because they are semi-digested. The earthworm species viz., *Eisenia foetida, Eudrillus eugineae, Amyanthes diffringens, Perionyx excavalus* etc. are some of the efficient species for making the compost.

Method

Structure: A manger like structure is constructed with bricks. The size of the structure is 90 cm in width, 50 cm in height and 6 m in length. The floor is made of cement plaster. A roof made of rice straw or thatch or C.G.I. sheets is erected. Direct sun light and rain should not fall on the compost beds. The earthworms like darkness and shade.

- Put a layer of dry soil or sand of 3 –5 cm thick on the floor of the tank.
- Break the spent beds and put in a layer of 20 cm thick over the dry soil.

- Put a 10 cm thick layer of cow dung or cow dung slurry over the spent mushroom substrate.
- Fill the compost beds with alternate layers of substrate and cow dung.
- Release the earthworms in the substrate @ 2 kg earthworm per ton of substrate.
- Cover the bed with wet gunny bags or straw.
- Sprinkle water so that the moisture content of the straw will be 70- 80%.
- It will take one month for decomposition of the materials. The spent beds are converted into vermicompost at a faster rate compared to other substrates. The reason is that the cellulosic material was already broken down by the cellulolytic enzymes released from the mycelium of mushroom.
- Stop watering when 90% of the substrate is decomposed. Maturity of the beds can be judged by seeing the granular form of the compost.

How to separate earthworms from the compost ?

Vermicompost is made into a heap. It is kept as such for 12 hours. All the worms will go down to the bottom. Thus the compost can be separated. Compost can also be separated by sieving.

The percentage of conversion of spent beds into vermicompost is 50. That is 100 kg of spent beds will yield 50 kg of vermicompost.

Nutrients in vermicompost made from spent beds

Nitrogen	1.85%
Phosphorus	0.90%
Potash	1.12%

This compost produced completely by organic means is extremely fertile and highly suitable for use in the cultivation of many crop plants, vegetables, flowering plants and shrubs. It can improve crop health and increase yield apart from improving the soil health.

Literature referred

Annonymous (1990). Recent Development in the Cultivation Technology of Edible Mushrooms. A compendium of lectures of Summer Institute, National Centre for Mushroom Research Research and Training (ICAR), Solan.

Atal, C.K., B.K. Bhat and T.N. Kaul (Eds.) (1978). Indian Mushroom Science-I. First National Symposium on Survey and Collection of Edible Mushrooms of India held at Regional Research Laboratory, Srinagar, Sept.22-24, 1976. Indo-American Literature House, Globe-Arizona, USA. pp. 531.

Bahl, Nita (2000). *Handbook on Mushroom* . New Delhi. Oxford & IBH Publishing Co. Pvt. Ltd., Oxford and New York. pp.166.

Bels-Koning, H.C. and W.M. van Kujik (1980). Mushroom Terms. Centre for Agricultural Publishing and Documentation, Wageningen. pp. 312.

Chadha, K.L. and S.R. Sharma (Eds.) (1995). Advances in Horticulture (Vol. 13, Mushroom). Malhotra Publishing House, New Delhi. pp. 649.

Chang, S.T. and W.A. Hayes (Eds.) (1978). The Biology and Cultivation of Edible Mushroom. Academic Press. New York. pp. 819.

Chang, T. and P.G. Miles (1993). Edible Mushrooms and Their Cultivation. CBS Publication, Delhi. pp.345.

Gokulapalan, C. and M.C. Nair (1990). Poisonous mushrooms. In Mushrooms. Technical Bulletin-17. M.C. Nair (Ed.). Kerala Agricultural University, Mannuthy. pp. 67-71.

Hard, M.E. (1961). The Mushroom. Hafner Publishing Company. New York. pp.609.

Jandaik, C. L. and S. Thinga (1980). Studies on cultivation and food value of *Macrolepiota procera*. *Mushroom Science* 11: 725-733.

Kapoor, J.N. (1978). Mushroom Cultivation. ICAR, New Delhi. pp.

Kibby, G. and S. Milne (1979). Mushroom and Toadstools, a field Guide. Oxford University Press, Oxford and New York. pp.255.

Pacioni, G. and G.H. Lincoff (1981). Simon and Schuster's guide to Mushrooms. Simon and Schuster Inc., New York and London. pp. 511.

Quimio, T.A. (Ed.) (2001). Workbook on Tropical Fungi, collection, isolation and identification. The Mycological Society of the Philippines, Inc.

Rindali, A. and V. Tyndalo (1974). The Complete Book of Mushrooms. Cresent Book, New York.

Singh, R.P. and H.S. Chaube (Eds.) (1995). Mushroom production technologies. G.B. Pant University of Agriculture and Technology, Pantnagar. pp. 104.

Smith, A.H. (1949). Mushrooms in their Natural Habitat. Hafner Press, a division of Macmillan Publishing Co., Inc., New York. pp. 626.

Smith, A.H. (1963). The Mushroom Hunters Field Guide. Ann Arbor. The University of Michigan Press. pp. 187.

Vijay, B. and H.S. Sohi (1987). Cultivation of oyster mushroom, *Pleurotus sajor-caju* (Fr.) Singer on chemically sterilized wheat straw. *Mush. J. Tropics* 7: 67-75.

Index